D0025409

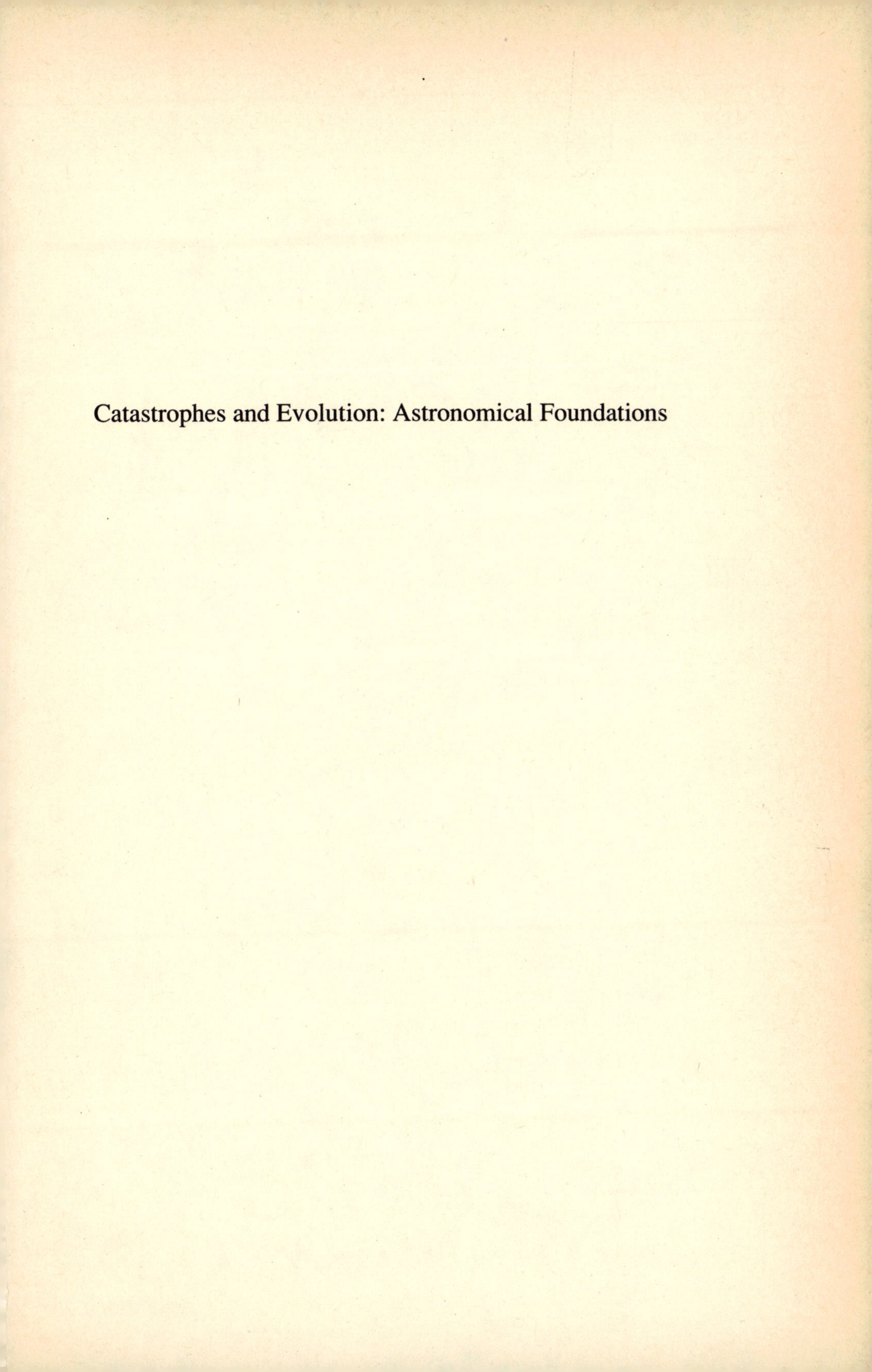

Catastrophes and Evolution: Astronomical Foundations

Catastrophes and Evolution:
Astronomical Foundations

The 1988 BAAS Mason Meeting
of the
Royal Astronomical Society
at Oxford, September 6th, 1988

with presentations by

Walter Alvarez, Edward Anders, M.E. Bailey,
S.V.M. Clube (editor), Kenneth M. Creer, R.D. Davies,
Richard A.F. Grieve, A. Hallam, W.M. Napier,
Duncan Olsson-Steel and A.W. Wolfendale.

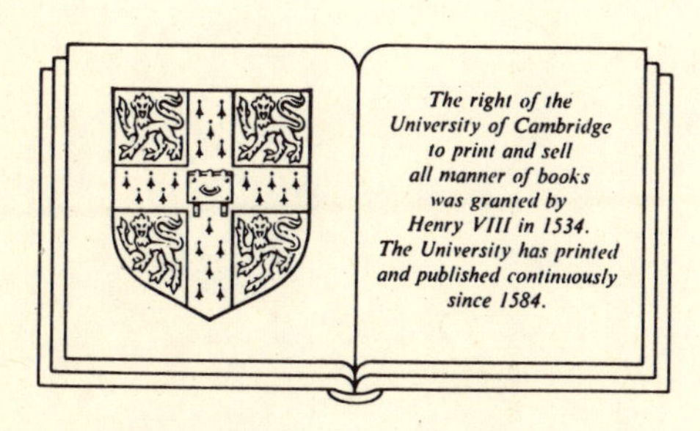

CAMBRIDGE UNIVERSITY PRESS

Cambridge
New York Port Chester
Melbourne Sydney

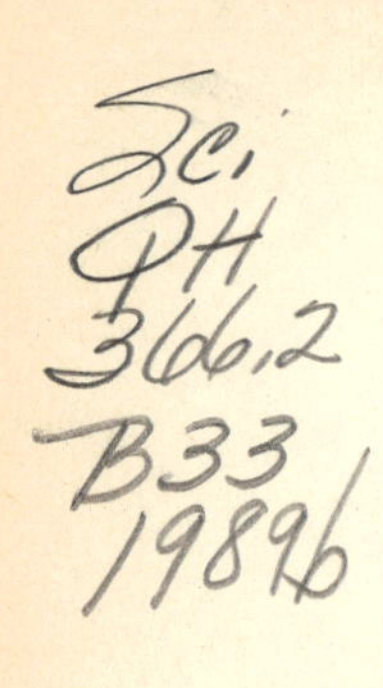

Published by the Press Syndicate of the University of Cambridge
The Pitt Building, Trumpington Street, Cambridge CB2 1RP
40 West 20th Street, New York, NY 10011, USA
10 Stamford Road, Oakleigh, Melbourne 3166, Australia

First published 1989

Printed in Great Britain at the
University Press, Cambridge

British Library Cataloguing in publication data available

Library of Congress cataloguing in publication data available

ISBN 0 521 37420 0

Dedication

In memory of Luis Alvarez,
who was not present

Contents

Preface

It is very nearly ten years since Luis Alvarez and his colleagues announced to an unsuspecting public that a huge blanket of cosmic material enveloped the Earth at the time of the famous dinosaur extinction. This finding, based upon trace element studies in the appropriate geological strata, notably of iridium, provided the first tangible evidence that the cosmos may be playing a dominant rôle in the evolution of life on Earth. Evidence for such a rôle knocks at one of the most established precepts of modern science, namely the essential truth of Darwinian theory which in turn presupposes a uniform terrestrial framework unaffected by its astronomical environment. Thus it is generally accepted at present that the evolution of life on Earth is characterised by a *steady* growth in the complexity of its lifeforms arising from the action of a purely intrinsic process, natural selection, in the presence of an essentially *steady* environment. It is hardly surprising that the world of science was plunged into considerable disarray when the disturbing announcement was made.

Astronomers in particular have since been making it their business to re-evaluate the circumstances under which the evolution of life on Earth probably occurs. The Royal Astronomical Society and the British Association for the Advancement of Science, as joint organisers of the astronomical Mason Conference at the latter's annual meetings, selected the 1988 Oxford Meeting of the British Association as a suitable occasion on which to review the progress that had since been made. Recognising the seminal rôle of the great debate at the 1860 Oxford Meeting of the British Association, in fixing the broad basis of modern ideas about the evolution of life on Earth, Luis Alvarez was very enthusiastic about a return to the Oxford setting under the auspices of the British Association, in order to retrace the scientific steps and perhaps discover how, during the intervening century or so, the fundamental error, for such it now seems to many to be, was made.

In the event, Luis Alvarez died less than a week before the Meeting took place, and it is everyone's loss that his considered reflections on a subject very much to his taste were never heard. This volume of papers, the contributions of others in the research field who were invited to the Meeting, may nevertheless be of interest to the community of readers whom he might have hoped to reach. Whilst it does not necessarily aspire to the same degree of robustness which he would have displayed in challenging the established thinking of the time, there was a widely expressed wish among the contributors that the volume be dedicated to the memory of an absent colleague who was largely responsible for initiating the line of enquiry which was the concern of the Meeting.

The nature of the Meeting was such that authors were encouraged to speculate, if they so wished, on the significance of recent trends. A degree of repetition was perhaps inevitable in the circumstances and it was expected that editorial scissors might have to be applied should the readers of this volume be unduly burdened with repeated evidence. As it turned out, the area of overlap was minimal and little adjustment was necessary. Where however a speaker chose subsequently to present a contribution in collaboration with other authors, the participating speaker alone has been headlined in the text.

The proceedings of the Meeting were also such as to be essentially complementary to those of the independently organised conference on global catastrophes in Earth history which was held during the following month in Snowbird, Utah. In practice, the emphasis of this particular conference differed considerably from that of the present meeting: for, whereas in the United States there is at present a tendency to see catastrophism mainly as an issue for geologists, most of the concern apparently being with large impacts whose existence and overall frequency are not really in doubt from the astronomical end, the tendency on the other side of the Atlantic, in Britain especially, is to see catastrophism more as a problem for astronomers — the concern being to ensure that we have a proper understanding of the rôle of encounters with small as well as with large bodies in arriving at a valid theory of evolution. The title of this volume, pointing as it does to the astronomical foundations of catastrophism, thus draws attention to an aspect of the subject which assumed particular prominence at the Meeting.

In the wake of the original announcement by Alvarez, it was soon recognised of course that large asteroid encounters and mass extinctions, though equally frequent and very naturally associated, happen too often for this particular combination of events to be the whole story as far as the heavy concentration of cosmic material at the dinosaur extinction was concerned. Astronomers therefore were exceptionally taxed by the problem of what extra could be involved. Two main ideas appear to have emerged and these receive a good deal of attention in the pages that follow. On the one hand there are those who appear still to regard the elements of Darwinian theory as substantially true whilst also arguing for occasional major upsets in evolution due to huge showers of comets from a possible inner extension of the Oort cometary cloud which some astronomers now believe to be present and to have been unfortunately overlooked in the past. On the other hand there are those who regard the Darwinian theory as very seriously diminished by the new finding whilst arguing instead for a virtually continuous contribution to evolution from frequent very large comets whose properties astronomers now recognise as not having been previously well understood and which may in fact be a prime source of asteroids. According to this line of investigation, large comets are natural repositories for the now very wide range of organic molecules observed in cosmic space and are therefore plausible sources of biogenically active units of life as well.

The choice between these alternatives is not without its piquancy. In the former case *homo sapiens* may have to wait millions of years before the Earth is again seriously disturbed by a comet shower. Evolution in the immediate future remains progressive therefore, underpinned by the gentler Darwinian process. In the latter case, a swarm of cosmic debris due to one giant comet, such as was responsible in the past for the dinosaur extinction when the Earth, exceptionally, penetrated its dense core, happens to be active at present and may even have been involved in close encounters of differing intensity only as recently as five thousand and two thousand years ago, the next serious incident being merely a thousand years or so ahead. Accordingly we should now be seeking the evidence in the ground for global fires and tidal floods on a rather more frequent timescale than heretofore supposed. Indeed, it seems that catastrophes may be as important for civilisation as they are now believed to be for evolution.

Several names should be mentioned here in connection with the preparation for the British Association/Royal Astronomical Society Meeting of which this is the partial record. First, the President of the Royal Astronomical Society, Professor Rod Davies (University of Manchester), and the Organising Secretary, Dr Carole Jordan (University of Oxford), who gave enthusiastic support and secured financial assistance beyond the norm on behalf of the Society. Second, Dr Frank Close (Rutherford-Appleton Laboratory) and Dr John Dainton (University of Liverpool) who, along with Dr Connie Martin of the British Association for the Advancement of Science, made the arrangements for the Meeting. Third, Professor Donald Blackwell (University of Oxford) who made available the facilities of the Department of Astrophysics at Oxford for the successful Workshop that was held in conjunction with the Meeting. Turning to the publication of the Proceedings, nothing would have happened without the enthusiasm and support of Dr Simon Mitton and Dr Caroline Roberts of the Cambridge University Press. The contributions of all these people are gratefully acknowledged. Last but by no means least, I wish to thank Dr David Petford and Miss Lesley Clarkson of the Department of Astrophysics at Oxford for patiently transforming the various manuscripts into the camera-ready form that follows and for a level of editorial assistance that essentially rendered the editor redundant.

S.V.M. Clube
Oxford, U.K.

July 1989

CATASTROPHES AND EVOLUTION.

THE 1988 BAAS MASON MEETING OF THE ROYAL ASTRONOMICAL SOCIETY AT OXFORD

R.D.Davies, President of the RAS

Nuffield Radio Astronomy Laboratories, Jodrell Bank, Macclesfield, Cheshire SK11 9DL, UK

Summary: Attention is drawn to the importance of the continuing "uniformitarianism" versus "catastrophism" debate. In particular, the significance of palaeontological and geological data in relation to astronomical periodicities in the terrestrial record is emphasized. Also, whilst there is little doubt that impacts are important so far as mass extinctions are concerned, the complexities of the terrestrial record and the various astronomical possibilities are such that much is still obscure and there is need for maintaining the dialogue between astronomers and Earth scientists to resolve the very fundamental issues at stake.

Introduction

A previous meeting of the British Association for the Advancement of Science (BAAS) held on the 30th June 1860 saw one of the most famous confrontations between science and entrenched attitudes. This was the debate between Thomas Huxley, a leading naturalist, and Samuel Wilberforce, Bishop of Oxford, on the subject of Darwinian evolution. The debate, held in the same building (the University Museum) as the present discussion meeting, was a confrontation between the Darwin-Wallace proposal that natural selection was a mechanism for the transmutation (modification) of species and the more conservative views of the time represented by the bishop. Charles Darwin's treatise *"Origin of the Species"*, published in November 1859, was based on 20 years of observation and provided the focus for the debate.

The subject of evolution still continues to attract wide attention by scientists and the public. Indeed this meeting on "Catastrophes and Evolution" has attracted the largest audience since the Mason meetings were introduced by the BAAS in 1983. Although the Huxley-Wilberforce encounter is often seen as the reaction of a cloistered ecclesiastical establishment to the radical hypotheses of science, the actual situation was far more complex. Thus the scientific establishment itself, whilst honouring Darwin the naturalist, drew back from acknowledging his "speculations" about evolution. Indeed, the citation for the Royal Society's award of the Copley Medal makes no mention of his theory of natural selection. Its wide acceptance in due course was gradual and even then not universal.

It is interesting that our discussion meeting today has as its core a similar emphasis to that raised by the *"Origin of the Species"*, namely the question of a gradual versus a sudden evolution of species. Naturalists (zoologists and botanists) immediately identified with the problems of the debate on evolution. Geologists retrospectively have seen the debate as winning the "uniformitarianism" versus "catastrophism" argument. Astronomers and physicists had no say in the argument and so it can only be assumed that they too condoned the uniformitarian position.

Prior to the publication of the *"Origin of the Species"* there had been a re-emergence of scientific catastrophism in the period 1800-1830 led by the more secularly inclined French, including such names as Biot, Cuvier and Laplace. Their arguments were both astronomical and geological. Contemporary British thought on the other hand was not so secular, most of the support for catastrophism at the time coming from learned clergymen, of whom Buckland at Oxford was a leading figure, through the subject's ready association with biblical catastrophism (so-called diluvialism). However, by the 1840's and 1850's something of a scientific reaction had set in amongst British geologists; and Lyell was foremost in arguing for the long-term action of present-day forces to explain geological evolution: indeed, the terms "uniformitarianism" and "catastrophism" were only coined by Whewell in 1837. Moreover the appearance of the *"Origin"* in 1859 was conceived as showing that natural selection in the biological sphere was a slowly acting uniformitarian force which was thus interpreted as setting religion and catastrophism at loggerheads with science and uniformitarianism.

This particular polarization of views is no longer upheld and today's discussion focusses on the possible role of catastrophes in the evolutionary process. The astronomer, the geophysicist and the geologist all see the potential for catastrophic situations at the surface of the Earth which can bring sudden changes to the biological populations. This can be the death or near-extinction of an existing population on one hand or the creation of a window of opportunity for a new population on the other. It is often said, rightly or wrongly, that the extinction of the dinosaurs at the Cretaceous-Tertiary boundary led to the emergence of man through the opportunities given to mammals by the demise of the dinosaurs.

The great extinctions in the Earth's biological record suggest the possible action of catastrophic events. Nevertheless, the palaeontological record is not incisive in providing an actual timescale for the extinction process; it may take a few years or a few million years. The mechanisms proposed also have timescales in this range. On the longer timescales, plate tectonics with its associated changes in ocean levels or in a long epoch of volcanic activity may be regarded as gradualist rather than catastrophic; yet the end effect on a species which cannot cope on this timescale will be terminal. On the shorter timescale, the planetary astronomer is familiar with comets and their associated meteor streams which may impact the Earth. Likewise some of the asteroids originating in the Mars-Jupiter belt also pass near the Earth and have a finite probability of colliding with it. The stellar astronomer would look to the supernova phenomenon, the ultimate collapse of a star followed by the ejection of a large fraction of its mass; the radiant energy from such explosions which occur every 30 years or so in the Milky Way would, if sufficiently close, provide major damage to life on Earth. Such an event is rare; only one would have happened in the lifetime of the Earth within a distance of a parsec where the radiation would approximately equal the Sun's radiant output for a brief period. Collisions with minor members of the Solar System, and large enough to have significant effects upon evolution, are much more common.

It is fitting at this point to pay tribute to one of those who in the present decade has given a major new stimulus to catastrophism, namely Luis Alvarez who died on 31 August 1988, only a week before this meeting. Alvarez, with his son Walter and collaborators, pursued with great enthusiasm the question of the mass extinction at the Cretaceous-Tertiary boundary, which they proposed was the result of an impact by a very large iron meteorite. Indeed, on drawing the threads of evidence together, he became utterly convinced of this explanation. In responding to the invitation by Victor Clube, the organizer of this meeting, to give the introductory lecture Luis Alvarez wrote "I am of course familiar with the famous Oxford meeting of the British Association, in June 1860, in which Huxley chopped up Bishop Wilberforce. In view of that history, plus the evidence from the "shocked quartz", I don't see how you'll be able to find someone "on the other side of the controversy" who will be willing to stand up and be counted – even though I can assure you that I won't try to emulate Huxley. My own view is that there is no longer any controversy except over the possible periodicity of the impacts, in which I am one of the few believers." It is fitting that Walter Alvarez has been able to attend and give the introductory lecture.

I will outline below some of the issues which need to be addressed if events in the palaeontological and geological records are to be expained in terms of astronomical phenomena.

The palaeontological/geological record

The geological time frame was historically specified in terms of the fossil content of the Earth's rocks. Eras were identified in order from the earliest, the Cambrian, to the most recent, the Tertiary. A quantitative timescale was later established through radioactive dating. The palaeontological record follows the geological eras so that the end of a geological period generally corresponds to a substantial extinction of species (see Figure 1).

Modern palaeontological data bases can be used to provide a reliable indication of the history of extinctions in the last half billion years. In the biological classification, species are grouped in genera which in turn are grouped in families. The analysis of extinctions is best made in terms of genera, of which some 25 thousand are known and 20 thousand are extinct. A graph of the rate of extinctions over time (Figure 1) plots the course of evolution over the last 500 million

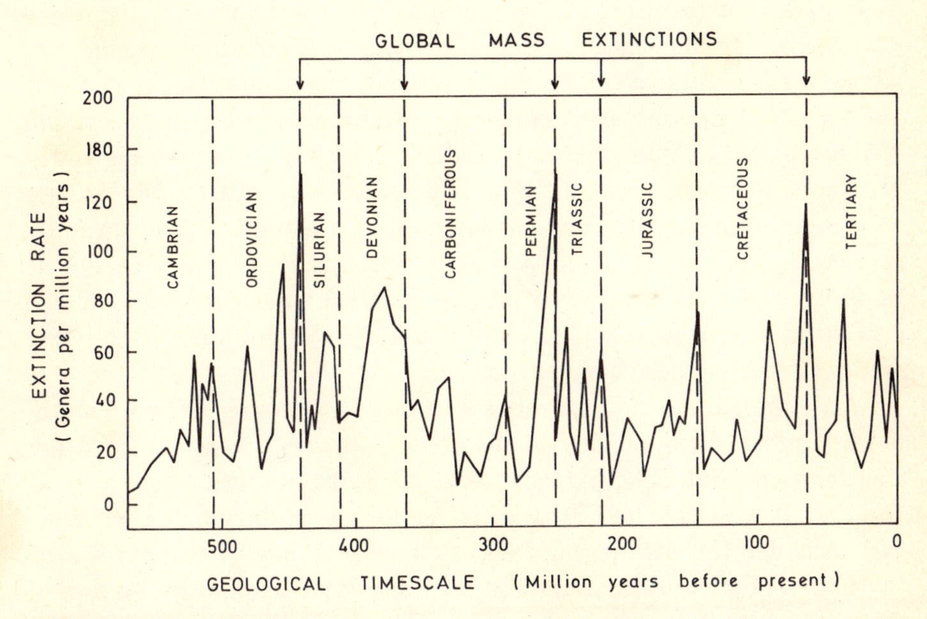

Figure 1. The extinction rate of genera per million years over the previous 550 million years (after Sepkoski).:

years. There is a background of 20-30 genera extinctions every million years above which rise peaks that are contenders for being major events in the Earth's history. Of particular interest are the mass extinctions, those which are world-wide, of considerable magnitude, affecting a range of families and occurring over a relatively short time. Five such mass extinctions are generally recognized.

The earliest identified mass extinction was at the end of the Ordovician, some 440 million years ago. Fifty-seven percent of marine invertebrate genera succumbed. Next came the end-Devonian extinction 370 million years ago; it is characterized by the vast size of the biomass which died out. The third and greatest mass extinction was the end-Permian in which 96 percent of marine invertebrate species died out in a few million years. This may have resulted in part from the loss of shallow coastal waters due to tectonic plate movements. The fourth was the end-Triassic 210 million years ago which saw the end of half the genera of marine invertebrates.

The final mass extinction, and the one most amenable to detailed study, occurred 65 million years ago at the Cretaceous-Tertiary boundary. Although this is famous for the demise of the huge dinosaurs, half the genera of that time perished, including some of the smallest such as microscopic floating animals and plants. However at the same time, land plants, crocodiles, snakes and mammals survived. A remarkable feature of this boundary layer has been the discovery of excessive amounts of iridium scattered over widely spaced sites around the Earth. Iridium, a platinum group member, is normally present in the core of the Earth but not at the surface where it is under-represented relative to normal cosmic abundances This is the evidence brought forward by Alvarez *et al.* (1980) for an impact origin of the Cretaceous-Tertiary extinction.

An important desideratum for an understanding of the role of catastrophism in evolution is the timescale over which a global extinction occurred. The palaeontological record is critical for this discussion. If we take the Cretaceous-Tertiary boundary as an example and consider the evidence from the disappearance of the dinosaurs, the bones of such large creatures can only give a rough indication of the timescale of the extinction — perhaps a million years. On the other hand a study of the distribution through sedimentary layers of microscopic life forms such as marine foraminifera can sharpen up the resolution to a precision approaching 10,000 years at the Cretaceous-Tertiary boundary. This is probably the limit for the palaeontological record. The ability to distinguish timescales of a few years which would be the extinction period for astronomical impact scenarios is far beyond present techniques. The geochemical data suggest a very rapid ($\leq$100 years) deposition of the anomalous elements at the boundary. Nevertheless, if the biological record indicates that these events occur on timescales of 10^4 rather than 10^6 years, the probability of a catastrophic cause is greatly increased.

It is sometimes argued that the effect of astronomical catastrophes may have only a modest effect in relation to terrestrial factors such as climate or sea-level changes. The latter may have a dominant role over long periods such as the extended Devonian extinction. At some point the astronomical catastrophe may then bring the *coup de grâce* to a weakened population. On the other hand, there are others who would argue for a more fundamental but still largely unrecognised role for astronomical catastrophes; and that these determine the course of terrestrial events on almost every conceivable timescale upwards of a thousand years.

Impacts as the cause of extinctions

Clearly, impacts of large objects on the Earth are likely to produce significant effects on terrestrial biological systems. The rate of infall of meteors, identified as the debris of comets, is well-documented. At least fifty tonnes of meteoric material enter the Earth's atmosphere

	Collisions per year		
Diameter (km)	0.1	1.0	10
Comets	$3.3\text{x}10^{-5}$	$2.5\text{x}10^{-7}$	$2.0\text{x}10^{-9}$
Mars asteroids	$1.0\text{x}10^{-6}$	$1.6\text{x}10^{-8}$	$3.7\text{x}10^{-10}$
Apollo group	$5.6\text{x}10^{-5}$	$1.1\text{x}10^{-7}$	$2.5\text{x}10^{-10}$
Sum of 3 contributions	$9.0\text{x}10^{-5}$	$3.8\text{x}10^{-7}$	$2.6\text{x}10^{-9}$
Interval (yr)	$1.1\text{x}10^{4}$	$2.6\text{x}10^{6}$	$4\text{x}10^{8}$

Table 1. The frequency of catastrophic collisions with celestial bodies; following Öpik 1958.:

every day. However, rare single bodies of this mass are also occasionally seen burning up in the atmosphere whilst the most massive actually seen to fall to the Earth (meteorites) are only a few tonnes in mass. The numbers of larger projectiles which are needed to produce significant extinctions require extrapolation both from these data and observations of the small body population of the Solar System.

As an example of the projections which can be made from observations of the minor members of the Solar System, I take the calculations made by Öpik (1958). He considered the contributions of comets, Mars-crossing asteroids and Apollo Earth-crossing asteroids for each of which he assumed a mass distribution based on available observational data. Table 1 gives the number of collisions per year for objects with diameters of 0.1, 1.0 and 10 kms. It can be seen that the interval between collisions for all the 0.1 km objects is $\sim 10^4$ years, rising to 400 million years for 10 km objects. Modern calculations would generally alter these numbers by at least a factor of $\sim 5 - 10$ and in the sense of increasing the collision rate (Clube & Napier 1982). Even so, Öpik was led to emphasize that life more than 500 million years ago may have been greatly hampered by catastrophic collisions arising from the large number of planetesimals in the past.

An impact may produce a whole suite of effects which could lead to a biological extinction. In the immediate vicinity of the impact, a blast wave is lethal over thousands of kilometres, ejected material may cover and destroy life, while further afield gigantic ocean waves will flush out the surrounding land. Sufficiently large impacts may trigger extensive volcanic activity in the Earth's crust. The greatest effect of the impact on the Earth's biology may be the ejection of dust into the atmosphere, forming a large reflective blanket around the Earth which will cause a world-wide drop in surface temperature sufficient to kill life forms directly or seriously disrupt food chains.

It is estimated that planetary bodies $\sim$ 10 km in diameter (with a mass of 10^{18}–10^{19} grams) striking the Earth with an energy of 10^{29}–10^{30} ergs are required to cause a global extinction. Such objects will leave a crater $\sim$ 100–200 km in diameter. Modern surveys of the Earth's surface morphology from spacecraft, supplemented with ground-based exploration, have discovered many tens of crater structures more than 10 km in diameter. Geological surveys of the crater boundaries confirm that they are impact features. Amongst the largest is the Manicougan crater in Canada 100 km in diameter which shows the fracture structure characteristic of an impact shock. This impact is on the lower limit of producing the energy required for the Cretaceous-Tertiary event ($\sim 10^{30}$ erg). A larger feature 300 km in diameter covering western Czechoslovakia was identified on images from the Meteosat 2 weather satellite. This would probably have the necessary energy; however it has still to be proven that it is an impact crater. Since two thirds of the Earth's surface is covered by sea or ice, there are many impact craters still to be discovered. It would seem not to be too difficult to identify enough large impact craters to give the mass extinction rate of approximately one per 10^8 years required by the palaeontological record.

Periodicity in the terrestrial record

An examination of the peaks in the pattern of extinctions over the last 550 million years suggest a periodicity of a few tens of millions of years. A more quantitative analysis of this data base has led some investigators to identify a period for extinctions of 26 million years; this is not a clearly defined period and may well be no more than a feature of the random noise of the time sequence. The problems in dating the geological series confuse the search for periodicity; time warps are inadvertently introduced which can smear out an intrinsic periodicity or artificially enhance a non-period.

Some credence has been placed in this estimated period because similar timescales are found in other global terrestrial phenomena. For example, 30 million years has been claimed for the period of the terrestrial cratering cycle while a period for volcanism of 15 or 30 million years has been suggested. The frequency of the Earth's magnetic field reversals also vary on these timescales; a period as well-defined as 14 ± 2 million years has been deduced. Whether these periodicities are reflections of some precise (astronomical) periodicity or the weakly-defined timescale of episodic events induced in some other way is not clear from the data; the latter is the weaker hypothesis and is thus more probable!

Several attempts have been made to find astronomical timescales of this order. So far as the more serious proposals are concerned, a period is first identified; however, as explained in the following examples, the actual effect on the Earth may be through a large number of smaller events spread over intervals of 10^5 yr which occur at the proposed astronomical period . The most basic proposal is that our Sun has a companion star, Nemesis, which orbits it with a period of $\sim$ 30 million years. When this star periodically approaches the Sun it disrupts the orbits of comets and produces a barrage of objects some of which collide with the Earth. No concrete evidence has been found for the existence of Nemesis. Another periodicity of the required magnitude is the oscillation time of the Sun through the galactic plane. The period of this z-motion is $\sim$ 60 million years; the relevant period for interactions is half this, the interval between galactic plane crossings. Every 30 million years or so, it is suggested, comets would be released from the Oort cloud far out on the periphery of the Solar System from which they would rain down on the Earth.

Studies of periodicities seem to raise more questions than they solve (Bailey, Wilkinson & Wolfendale 1987). With the present data it is unclear whether we are dealing with an accurate period or merely a characteristic timescale of episodic events. The general agreement between the timescale of extinctions and other terrestrial events including cratering is suggestive of a causal link but is not compelling. The physical mechanisms for producing the putative periodicities are, however, far from clear. For example, how can an impact giving a 10^{-9} change in the Earth's angular momentum produce a magnetic reversal when atmospheric drag already produces a factor of 10^{-7} and shows no associated magnetic effect?

Conclusion

It is a truism to say that the role of extraterrestrial (astronomical) impacts in biological evolution is complex. We can state categorically that global extinctions occur at a frequency of one per 50–100 million years (Figure 1) and that craters with diameters of 100–200 km are produced by extraterrestrial impacts on similar timescales. Whether these phenomena are cause and effect cannot yet be demonstrated with similar certainty. There is clearly room for a useful continuing dialogue on these matters between astronomers, geologists and palaeontologists. This meeting should encourage and facilitate this exchange. The subject is too exciting to leave dormant.

References

Alvarez, L.W., Alvarez, W., Asaro, W. & Michel, H.V., 1980. *Extra-terrestrial cause for the Cretaceous-Tertiary extinction.* Science, **208**, 1095-1108.

Clube, S.V.M. & Napier, W.M., 1982. *Spiral arms, comets and terrestrial catastrophism.* Q.Jl.R.Astron.Soc., **23**, 45.

Öpik, E.J., 1958. *On the catastrophic effects of collisions with celestial bodies.* Irish.Astr.J., **5**, 34.

Bailey, M.E., Wilkinson, D.A. & Wolfendale, A.W., 1987. *Can episodic comet showers explain the 30 Myr cyclicity in the terrestrial record?* Mon.Not.R.astr.Soc., **227**, 863.

UNIFORMITARIANISM AND THE RESPONSE OF EARTH SCIENTISTS TO THE THEORY OF IMPACT CRISES

Walter Alvarez[1], Thor Hansen[2], Piet Hut[3], Erle G. Kauffman[4] and Eugene M. Shoemaker[5]

[1]*Department of Geology and Geophysics, University of California, Berkeley, CA 94720, USA*
[2]*Department of Geology, Western Washington University, Bellingham, WA 98225, USA*
[3]*The Institute for Advanced Study, Princeton, NJ 08540, USA*
[4]*Department of Geological Sciences, University of Colorado, CB-250, Boulder, CO 80309, USA*
[5]*Branch of Astrogeology, U.S. Geological Survey, Flagstaff, AZ 86001, USA*

Summary. The doctrine of uniformitarianism strongly influences the way Earth scientists view the evolution of this planet, through a tradition which uses the modern world as a model for the past, assumes gradualistic changes, and shuns catastrophic explanations. Yet Gould's analysis of uniformitarianism shows that it is a confused mixture of two ideas. One of these ideas, "methodological uniformitarianism" is merely a reformulation of the basic assumption of scientific methodology. The other idea, "substantive uniformitarianism", or gradualism, is simply wrong. Internally consistent evidence now supports a temporal correlation of large-body impact with the mass extinction at the Cretaceous-Tertiary boundary. The past rate of large impacts on the Earth is in good agreement with the rate predicted from observations of orbiting objects. Large-body impacts are not *deus ex machina* explanations; they are inevitabilities. Yet because of the influence of uniformitarianism, many geologists and paleontologists prefer to explain mass extinctions by gradualistic mechanisms which require unlikely combinations of unrelated causal events. Earth science is now at a point where it can no longer afford to be shackled by a dogma of the nineteenth century. Although many Earth processes may in fact be gradualistic, others definitely are not. Strict uniformitarianism should be relegated to the status of a corollary to Occam's razor, and we should be prepared to accept the conclusions to which our evidence drives us.

The content of uniformitarianism

Since 1980, a great deal of physical and chemical evidence has been found to support the hypothesis that a large extraterrestrial body collided with the Earth at the time of the Cretaceous-Tertiary boundary, about 65 million years ago, and was coincident with widespread biological mass extinction at that time (Alvarez *et al.* 1980, Ganapathy 1980, Smit & Hertogen 1980, Kyte *et al.* 1980, Smit & Klaver 1981, Orth *et al.* 1981, Alvarez 1983, Luck & Turekian 1983, Montanari *et al.* 1983, Bohor *et al.* 1984, Alvarez 1986, Raup 1986, Hsu 1986, Alvarez 1987, Izett 1987, Muller 1988).

In the past three years, several articles (Sloan *et al.* 1986, Patrusky 1986-1987, Archibald 1987, Courtillot & Cisowski 1987, Hallam 1987, Officer *et al.* 1987, Crocket *et al.* 1988) have presented objections to the impact hypothesis. These recent articles favour the view that the Cretaceous-Tertiary mass extinction was not sudden, and was the result either of gradual changes in sea-level, ocean chemistry or climate, or an unusual pulse of volcanism. Although one or another of the present authors could argue on technical grounds with the data and conclusions in these papers, we think it may be more interesting to view these articles in perspective, as a case study in the philosophy of Earth history.

A careful reading of the anti-impact articles shows that they contain only weak arguments which purport to contradict the impact explanation for the terminal-Cretaceous mass extinction. The logic underlying the papers is, rather, that it is simply not *necessary* to invoke an impact, because the relevant physical, chemical, and paleontological data can also be explained by other phenomena, such as volcanism or a sea-level fall.

Why should the view that there is no *necessity* to invoke an impact carry any weight? Most Earth scientists will recognize the attempt to minimize the role of impacts in Earth processes as a manifestation of the doctrine of uniformitarianism. This term refers to a time-honoured but vaguely defined view that the present is the key to the past and that explanations of Earth history by gradual processes are preferable to explanations invoking sudden, and typically violent, processes. Gould (1965) has analyzed the intellectual content of uniformitarianism and has shown that it has two main formulations: "substantive uniformitarianism" is the notion that no geologic process has ever proceeded at a different rate in the past than it does now (a clearly false assumption), whereas "methodological uniformitarianism" is the refusal to accept miraculous explanations (an unnecessary admonition at this stage in the development of science).

It is a widespread view among geologists that uniformitarianism, as developed by Charles Lyell in the first half of the nineteenth century, provided an essential antidote to biblically inspired *ad hoc* catastrophism in the eighteenth and nineteenth centuries. However, Gould (1984, 1987) has also shown that this interpretation does injustice to the scientific catastrophists. He also points out that Lyell used the rhetorical trick of giving the same name to the very different concepts of substantive and methodological uniformitarianism, in order to push a rigidly gradualistic view of geological processes which can now be seen to be far from correct. Lyell was successful and uniformitarianism has subsequently been passed along from generation to generation as a cultural heritage of geology and paleontology. In our view, the uncritical acceptance of the doctrine now interferes with the rational development of the Earth sciences. Shea (1982) has shown in detail how this interference operates, in an essay entitled "Twelve fallacies of uniformitarianism".

Thus, methodological uniformitarianism, in its modern form, basically boils down to Occam's Razor (Shea 1982). It plays a useful role in the conservative approach scientists must take to unorthodox ideas, many of which will succumb to the testing process. But fallacies embedded in Lyellian uniformitarianism have delayed the acceptance of important advances in geology, particularly with regard to the role of catastrophic processes. One example is the case of the 'catastrophic flood' hypothesis for the origin of the Channeled Scablands of eastern Washington, proposed long ago by Bretz (1923), and not accepted until the 1950s (Gould 1984).

Catastrophic impacts : a geologic process

At the present time, the influence of substantive uniformitarianism is seen in the reluctance of many geologists to accept impact as a significant and inevitable geologic process. Impact structures on the Earth have commonly been attributed to unexplained "cryptoexplosions", and they still seem to be of little interest to a large segment of the general geological community. As one of us has written, "Most geologists just don't like the idea of stones the size of hills or small mountains falling out of the sky" (Shoemaker 1984).

Similarly, the reality of mass extinctions is contested by a few scientists who, following Lyell's view (Gould 1987), interpret apparently abrupt evolutionary events as evidence for gaps in the stratigraphic record. Thus it is little wonder that attributing a mass extinction to an impact provokes discomfort among many Earth scientists, and a search for alternative explanations is a natural response. If the heritage of uniformitarianism leads one to doubt the importance of impacts, one's scientific response to the evidence for a major impact at the Cretaceous-Tertiary boundary must be to show (1) that the evidence does not fit an impact scenario, and/or (2) that it does fit some other cause or causes.

Since no one has maintained that the primary lines of evidence (*i.e.* the anomalous iridium and other noble metals occurring with chondritic or Solar System abundance ratios, microspherules, shocked minerals and lithic fragments, osmium isotopic ratios, and world-wide distribution of at least the better studied of these features) are incompatible with a major impact, skeptics have argued that deposition of these features continued too long to have been due to an "instantaneous" impact event. At one time it was argued (Officer &

Drake 1983, Payne *et al.* 1983) that the Cretaceous-Tertiary boundary iridium anomaly was deposited in some areas during a time of normal geomagnetic polarity, and thus could not be synchronous with the iridium deposition in other areas which are well documented as having occurred during a time of reversed polarity. But after the studies suggesting occurrence of the iridium anomaly in a normal–polarity zone were shown to have been incorrect (Alvarez *et al.* 1984, Butler & Lindsay 1985, Shoemaker *et al.* 1987), that argument disappeared. The approach now (Courtillot & Cisowski 1987, Officer *et al.* 1987) is to argue that anomalous iridium deposition continued for as much as 10^4–10^5 years, within the 500,000–year reversed polarity interval 29R, which contains the extinction event. This argument has been supported by citing those stratigraphic sections in which anomalous iridium is spread over the greatest stratigraphic interval. However, spreading out of a narrow peak into a broad one is a natural effect of sediment disturbance by burrowing organisms, redeposition by currents, and chemical remobilization. We argue that it is more difficult to concentrate an originally broad distribution into a narrow spike than it is to smear out a sharp one.

The second necessity for those arguing against an impact is to show that the evidence is compatible with some other mechanism. For example, strongly shocked clasts of quartz, feldspar, quartzose sedimentary rocks and granite have been found in the Cretaceous-Tertiary boundary clay (Bohor *et al.* 1984, Bohor *et al.* 1987, Izett & Bohor 1987). Individual quartz grains have as many as seven sets of well-developed shock lamellae, are indistinguishable from shocked quartz found at known impact craters (French & Short 1968), and represent one of the strongest lines of evidence for an impact. In a recent paper, Carter *et al.* (1986) claim to have found shocked quartz grains in volcanic ejecta. However, the geologists who have studied the shocked minerals and rock clasts from the Cretaceous-Tertiary boundary layer (Bohor *et al.* 1987; Izett & Bohor 1987) have shown that these grains are very different from the quartz grains found in volcanic rocks by Carter. The occurrence of the shocked lithic fragments in the boundary layer, in particular, argues against a volcanic origin. The supposed shocked quartz grains from volcanic ejecta contain only single sets of lamellae, a feature also found in quartz which has undergone slow, tectonic deformation. Even Carter appears to agree that the supposedly shocked quartz he has studied from volcanic ejecta differs from the quartz of the Cretaceous-Tertiary boundary and from impact craters, which all workers agree have been shocked (Kerr 1987).

A problem faced by the authors contesting the hypothesis of a major impact at the Cretaceous-Tertiary boundary is that even if a non-impact explanation can be found for each line of critical evidence, the same non-impact explanation does not seem to work for all of them (Alvarez 1986). To explain all the features of the Cretaceous-Tertiary boundary, Officer *et al.* (1987) requires two different kinds of volcanism. Violent, explosive volcanism is offered to explain the shocked quartz although, as noted above, it is probably an insufficient explanation. A great outpouring of basaltic lava is offered to explain the microspherules and iridium anomaly (although measured basalts are much lower in iridium than the peak values observed in the Cretaceous-Tertiary boundary). Major violent eruptions are produced by magmas high in silica and extremely low in iridium, whereas the highest iridium-bearing magmas are basaltic. Basalts have low silica content, are seldom quartz bearing, and are not known to give rise to cataclysmic eruptions of the type associated with highly silicic magmas. There is virtually no geographic overlap between flood basalts and major violently eruptive silicic volcanic centers. So Officer *et al.* (1987) propose a general increase of world volcanism at the time of the Cretaceous-Tertiary boundary, citing a few local areas where such an increase is inferred. This increase is not general, however, as Kauffman (1985), for example, has shown that the volume of volcanic ash deposited in the Western Interior Basin of North America in the Maastrichtian (latest Cretaceous) is the lowest recorded for any part of the Cretaceous in this area.

Further compounding the difficulties with incompatibility of causes, Officer *et al.* (1987) attribute much of the biological extinction event to a sea-level fall at the time of the suggested volcanic maximum. However, a global volcanic pulse suggests rapid mantle convection, which implies rapid sea-floor spreading, which in turn implies an increased volume of the mid-ocean ridges because of the lower density of the hot material they incorporate. This should yield a sea-level rise, not a fall (Pitman 1978). We conclude that the suggested styles of volcanism are incompatible in a single volcanic region, and that a world-wide pulse of both kinds of volcanism is neither documented nor compatible with a sea-level fall.

Eustatic sea-level fall and large-scale volcanism would be expected to continue over a substantial time interval, probably exceeding a million years, and to have produced selective extinctions, perhaps concentrated in the regions of active volcanism. However, the main extinction event, which coincides with the iridium anomaly at the Cretaceous-Tertiary boundary, was abrupt (probably 1-100 years) and affected ecologically and genetically diverse taxa (*i.e.*calcareous and siliceous plankton, diverse tropical to temperate molluscs, brachiopods, bryozoa, foraminifera, vertebrates and angiosperms), including groups at the evolutionary peak of their development (planktic foraminifera, nannoplankton and shallow-water molluscs). The rates and patterns of extinction across the Cretaceous-Tertiary boundary do not fit the predictions of the volcanic or sea-level mechanisms.

In essence, Officer *et al.* (1987) do not argue strongly against an impact, but they consider it unnecessary. Their proposed alternative, however, appears to us to be an unlikely combination of causes not known to have occurred together in the Earth's past. On the other hand, the impact hypothesis is compatible with all the known data, and impacts are events that are known to occur – impact craters are found on all the rocky planets and satellites, including the Earth and the Moon. About 80 Earth-crossing asteroids have been discovered telescopically, including objects up to about 10 km in diameter – the size of the proposed Cretaceous-Tertiary impactor (Wetherill & Shoemaker 1982, Shoemaker 1983, Shoemaker & Wolfe 1986, Shoemaker *et al.* 1988). Our present catalogue of impacting bodies, moreover, is very incomplete; on the basis of the rate of discovery in systematic surveys, Shoemaker *et al.* (1979) estimated that the population of Earth-crossing asteroids larger than 1 km in diameter is about 20 times greater than the set now known. The total rate of collision of these objects with the Earth is consistent with the geologic record of impact cratering over the last 120 Myr on the carefully studied shield areas of North America and Europe (Grieve 1984). Comet nuclei must also be included in the list of known impactors, and the recent spacecraft missions to Halley's Comet (Keller *et al.* 1986, Sagdeev *et al.* 1986) demonstrated once and for all that solid nuclei up to 10 km in diameter, and probably much larger, occur on Earth-crossing orbits. We cannot continue to exclude large-body impacts from the list of known geological processes.

Conclusion

Echoing the view of Goodman (1967) and Shea (1982), one of us has argued that "Perhaps it is time to recast uniformitarianism as merely a sort of corollary to Occam's razor, to the effect that if a set of geological data can be explained by common, gradual, well-known processes, that should be the explanation of choice, but that when the evidence strongly supports a more sudden, violent event, we will go where the evidence leads us" (Alvarez 1986). We submit that impact and the resulting environmental disturbances provide a far more likely causal mechanism for the Cretaceous-Tertiary mass extinction than a combination of apparently incompatible geological events. The impact theory is in accord both with Occam's Razor (uniformitarianism in its modern form) and with the general spirit of uniformitarianism, understood in the sense that processes such as impacts, which are known to occur or are statistically predictable, are better explanations for events in Earth history than are unknown mechanisms and *ad hoc* combinations of incompatible events.

Although we are persuaded that a large-body impact played a central role in the terminal-Cretaceous mass extinction, we are also of the opinion that the evidence argues for a more complicated story than a single large impact causing a single great extinction. Students of large-body impacts and of mass extinction have found both reason to predict, and evidence to support, multiple impact events, complex environmental effects, and a complicated fabric of extinction which differs from one mass extinction to the next (Porch-Nielsen *et al.* 1982, Lewis *et al.* 1982, Muller 1985, Glass *et al.* 1985, Kauffman 1986, Muller & Morris 1986, Shoemaker & Wolfe 1986, Hut *et al.* 1987, Keller *et al.* 1987). We believe that science has much more to gain from an open-minded exploration of the evidence for catastrophic events in Earth history than from a continuing insistence on fitting all our data into a nineteenth-century uniformitarian viewpoint.

Acknowledgement

This paper began with discussions among the authors during a meeting hosted by the Institute of Advanced Studies in Princeton, in the spring of 1987.

References

Alvarez, L.W., 1983 *Experimental evidence that an asteroid impact led to the extinction of many species 65 million years ago:* Proc. Natl. Acad. Sci. USA, **80**, 627-642.

Alvarez, L.W., 1987 *Mass extinctions caused by large bolide impacts:* Physics Today, **40**, 24-33.

Alvarez, L.W., Alvarez, W., Asaro, F. & Michel, H. V., 1980, *Extraterrestrial cause for the Cretaceous-Tertiary extinction:* Science, **208**, 1095-1108.

Alvarez, W., 1986 *Toward a theory of impact crises:* EOS, **67**, 649-658.

Alvarez, W., Alvarez, L.W., Asaro, F. & Michel, H.V., 1984, *The end of the Cretaceous: sharp boundary or gradual transition?* Science, **223**, 1183-1186.

Archibald, J.D., 1987 *Stepwise and non-catastrophic Late Cretaceous terrestrial extinctions in the Western Interior of North America: testing observations in the context of an historical science:* Memoires de la Societe Geologique de France, N.S., **150**, 45-52.

Bohor, B.F., Foord, E.E., Modreski, P.J. & Triplehorn, D. M., 1984 *Mineralogic evidence for an impact event at the Cretaceous-Tertiary boundary:* Science, **224**, 867-869.

Bohor, B.F., Modreski, P.J. & Foord, E.E., 1987 *Shocked quartz in the Cretaceous-Tertiary boundary clays: Evidence for a global distribution:* Science, **236**, 705-709.

Bretz, J.H., 1923 *The channeled scablands of the Columbia Plateau:* Jour. Geol., **31**, 617-649.

Butler, R.F. & Lindsay, E.H., 1985 *Mineralogy of magnetic minerals and revised magnetic polarity stratigraphy of continental sediments, San Juan Basin, New Mexico.* Journal of Geology, **93**, 535-554.

Carter, N.L., Officer, C.B., Chesner, C.A. & Rose, W.I., 1986 *The Dynamic deformation of volcanic ejecta from the Toba caldera: Possible relevance to Cretaceous/Tertiary boundary phenomena:* Geology, **14**, 380-383.

Courtillot, V.E. & Cisowski, S., 1987, *The Cretaceous-Tertiary boundary events: external or internal causes?*: EOS, **68**, 193-200.

Crocket, J.H., Officer, C.B., Wezel, F.C. & Johnson, G.D., 1988 *Distribution of noble metal across the Cretaceous/ Tertiary boundary at Gubbio, Italy: Iridium variation as constraint on the duration and nature of Cretaceous Tertiary boundary events*: Geology, **16**, 77-80.

French, B.M. & Short, N.M., 1968 SHOCK METAMORPHISM OF NATURAL MATERIALS: Baltimore, Mono Book Corp.

Ganapathy, R., 1980 *A major meteorite impact on Earth 65 million years ago: Evidence from the Cretaceous-Tertiary boundary clay:* Science, **209**, 921-923.

Glass, B.P., Burns, C.A., Crosbie, J.R. & DuBois, D.L., 1985, *Late Eocene North American microtektites and clinopyroxene-bearing spherules*: Jour. Geophys. Res., **90**, D175-D196.

Goodman, N., 1967 *Uniformity and simplicity.* Geol. Soc. Amer. Special Paper, **89**, 93-99.

Gould, S.J., 1965 *Is Uniformitarianism necessary?* American Journal of Science, **263**, 223-228.

Gould, S.J., 1984 *Toward the vindictation of punctuational change.* In CATASTROPHISM AND EARTH HISTORY; THE NEW UNIFORMITARIANISM: Princeton University Press, 9-34.

Gould, S.J., 1987 TIME'S ARROW, TIME'S CYCLE: Cambridge, Mass., Harvard University Press.

Grieve, R.A.F., 1984 *The impact cratering rate of recent time*: Jour. Geophy. Res., **89**, B403-B408.

Hallam, A., 1987 *End-Cretaceous mass extinction event: argument for terrestrial casuation*: Science, **238**, 1237-1242.

Hsu, K.J., 1986, THE GREAT DYING. San Diego, Harcourt Brace Jovanovich.

Hut, P., Alvarez, W., Elder, W.P., Hansen, T., Kauffman, E.G., Keller, G., Shoemaker, E.M. & Weissman, P.R., 1987, *Comet showers as a cause of mass extinctions*: Nature, **329**, 118-126.

Izett, G.A., 1987 *The Cretaceous-Tertiary (K-T) boundary interval, Colorado and New Mexico, and its content of shock-metamorphosed minerals: implications concerning the K/T boundary impact-extinction theory*: United States Geological Survey Open File Report, **87-606**, 1-125.

Izett, G.A. & Bohor, B.F., 1987 *Comment on "Dynamic deformation of volcanic ejecta from Toba caldera: Possible relevance to Cretaceous/Tertiary boundary phenomena"*: Geology, **15**, 90.

Kauffman, E.G., 1985 *Cretaceous evolution of the Western Interior Basin of the United States*: Society of Economic Petrologists and Mineralogists Field Trip Guidebook, **4**, iv-xiii.

Kauffman, E.G., 1986 *High resolution event stratigraphy: regional and global Cretaceous bio-events*: Lecture Notes in Earth Sciences, Global Bio-Events, **8**, 279-335.

Keller, G., D'Hondt, S.L., Orth, C.J., Gilmore, J.S., Oliver, P.Q., Shoemaker, E.M. & Molina, E., 1987, *Late Eocene impact microspherules: stratigraphy, age, and geochemistry*: Meteoritics, **22**, 25-60.

Keller, H.U., Arpigny, C., Barbieri, C., Bonnet, R.M., Cazes, S., Coradini, M., Cosmovici, C.B., Delamere, W.A., Huebner, W.F., Hughes, D.W., Jamar, C., Malaise, D., Reitsema, H. J., Schmidt, H.U., Schmidt, W.K.H., Seige, P., Whipple, F. L. & Wilhelm, K., 1986 *First Halley multicolour camera imaging results from Giotto*: Nature, **321**, 320-326.

Kerr, R.A., 1987 *Asteroid impact gets more support*: Science, **236**, 666-668.

Kyte, F.T., Zhou, Z. & Wasson, J.T., 1980 *Siderophile-enriched sediments from the Cretaceous-Tertiary boundary*: Nature, **288**, 651-656.

Lewis, J.S., Watkins, G.H., Hartman, H. & Prinn, R.G., 1982 *Chemical consequences of major impact events on Earth.* Geol. Soc. Amer. Spec. Pap., **190**, 215-221.

Luck, J.M. & Turekian, K.K., 1983 *Osmium^{187}/Osmium^{186} in maganese nodules and the Cretaceous-Tertiary boundary.* Science, **222**, 613-615.

Montanari, A., Hay, R.L., Alvarez, W., Asaro, F., Michel, H.V.,Alvarez, L.W. & Smit, J., 1983 *Spheroids at the Cretaceous-Tertiary boundary are altered impact droplets of basaltic composition.* Geology, **11**, 668-671.

Muller, R.A., 1985 *Evidence for a solar companion star*, In THE SEARCH FOR EXTRATERRESTRIAL LIFE: RECENT DEVELOPMENTS. D. Reidel Publishing, 233-243.

Muller, R.A., 1988 NEMESIS. New York, Weidenfeld and Nicolson.

Muller, R.A. & Morris, D.E., 1986 *Geomagnetic reversals from impacts on the Earth.* Geophys. Res. Lett., **13**, 1177-1180.

Officer, C.B. & Drake, C.L., 1983 *The Cretaceous-Tertiary transition.* Science, **219**, 1383-1390.

Officer, C.B., Hallam, A., Drake, C.L. & Devine, J.D., 1987 *Late Cretaceous and paroxysmal Cretaceous-Tertiary extinctions.* Nature, **326**, 143-149.

Orth, C.J., Gilmore, J.S., Knight, J.D., Pillmore, C.L., Tschudy, R.H. & Fassett, J.E., 1981 *An iridium abundance anomaly at the palynological Cretaceous-Tertiary boundary in northern New Mexico.* Science, **214**, 1341- 1343.

Patrusky, B., 1986-1987 *Mass extinctions; the biological side*: Mosaic, **17**, 2-13.

Payne, M.A., Wolberg, D.L. & Hunt, A.A., 1983 *Magnetostratigraphy of a core from Raton Basin, New Mexico; implications for synchroneity of Cretaceous-Tertiary boundary events*: New Mexico Geology, **5**, 41-44.

Perch-Nielson, K, McKenzie, J. & He, Q., 1982 *Biostratigraphy and isotope stratigraphy and the 'catastrophic' extinction of calcareous nannoplankton at the Cretaceous/Tertiary boundary*: Geological Society of America Special Paper, **190**, 353-371.

Pitman, W., 1978 *The relationship between eustasy and the stratigraphic sequences of passive margins*: Geological Society of America Bulletin, **89**, 1389-1403.

Raup, D.M., 1986 THE NEMESIS AFFAIR: New York, W.W. Norton.

Sagdeev, R.Z., Blamont, J., Galeev, A.A., Moroz, V.I., Shapiro, V.D., Shevchenko, V.I. & Szego, K., 1986 *Vega spacecraft encounters with Comet Halley*: Nature, **321**, 259-262.

Shea, J.H., 1982 *Twelve fallacies of uniformitarianism*: Geology, **10**, 455-460.

Shoemaker, E.M., 1983 *Asteroid and comet bombardment of the Earth*: Ann. Rev. Earth Planet. Sci. **11**, 461-494.

Shoemaker, E.M., 1984 *The Acceptance of the G.K. Gilbert Award*: Geological Society of America Bulletin, **95**. 1001-1002.

Shoemaker, E.M., Pillmore, C.L. & Peacock, E.W., 1987 *Remanent magnetization of rocks; latest Cretaceous and earliest Tertiary age from drill core at York Canyon, New Mexico*: Geological Society of America special paper, **209**, 131-150.

Shoemaker, E.M., Williams, J.G., Helin, E.F. & Wolfe, R.F., 1979 *Earth-crossing asteroids: orbital classes, collision rates with Earth, and origin.* In ASTEROIDS: University of Arizona Press 253-282.

Shoemaker, E.M. & Wolfe, R.F., 1986 *Mass extinctions, crater ages, and comet showers*, in THE GALAXY AND THE SOLAR SYSTEM: Tucson, University of Arizona Press, 338-386.

Shoemaker, E.M., Shoemaker, C.S. & Wolfe, R.F., 1988 *Asteroid comet flux in the neighbourhood of the Earth.* Abstract volume for meeting on "GLOBAL CATASTROPHES IN EARTH HISTORY", Snowbird, Utah, 20-23 October 1988, Lunar and Planetary Institute, Houston, 174-176.

Sloan, R.E., Rigby, J.K., van Valen, L.M. & Gabriel, D., 1986 *Gradual dinosaur extinction and simultaneous ungulate radiation in the Hall Creek Formation*: Science, **232**, 629-633.

Smit, J. & Hertogen, J., 1980 *An extraterrestrial event at the Cretaceous-Tertiary boundary*: Nature, **285**, 198-200.

Smit, J. & Klaver, G., 1981 *Sanidine spherules at the Cretaceous-Tertiary boundary indicate a large impact event*: Nature, **292**, 47-49.

Wetherill, G.W. & Shoemaker, E.M., 1982, *Collision of astronomically observable bodies with the earth*: Geological Society of America Special Paper, **190**, 1-13.

CATASTROPHISM IN GEOLOGY

A. Hallam

School of Earth Sciences, University of Birmingham, Birmingham, U.K.

Summary: An historical survey is presented of ideas relating to the concept of "catastrophism" in geological studies during the last two centuries. It is noted in particular that the opposing concept of "uniformitarianism", in which there is assumed to have been an overall constancy of geological processes through time so that there is no need to invoke catastrophic change, is now considered rather extreme. During the nineteen sixties and seventies, a neo-catastrophist viewpoint has increasingly emerged in various branches of geology. Mass extinctions and their possible causes — bolide impact, climate, volcanism and sea-level change for example — are each considered in the context of this developing framework.

Catastrophism in the Nineteenth Century

Geology began to emerge as an independent science at the transition from the eighteenth to the nineteenth century. If prime credit can be directed to any one person in particular it was the German mineralogist Abraham Werner rather than the Scotsman James Hutton, despite the claims of generations of British geologists. Werner's "neptunean" system involved far more than the supposed aqueous origin of basalt; it was the foundation of historical geology and proved immensely influential in the early part of the nineteenth century (Laudan 1987). Only quite falsely, however, could his views be dubbed catastrophist, as was sometimes done after his death by polemicists determined to discredit his whole system. The true father of catastrophism was Georges Cuvier.

Cuvier was an outstandingly able comparative anatomist and palaeontologist who was a professor at the Natural History Museum in Paris, and held a dominant position in French science for several decades. By application of his so-called law of correlation of parts he had brilliant success in reconstructing the gross form and mode of life of many fossil vertebrates (Rudwick 1972) and was the first person to establish that certain species, such as the mammoth and giant sloth, had gone extinct. He also made substantial advances in stratigraphy, being one of the first, with his compatriot Brongniart, to pay close attention to the sequence of fossils, an approach that was to lead to a breakthrough in correlation methods. Cuvier and Brongniart made a detailed study of the strata and fossils of the Paris Basin, recognising a whole series of faunas back through the Tertiary to the Cretaceous Chalk, which appeared to end as abruptly as they began. They were able to demonstrate, by comparison with present-day deposits and organisms, that the Tertiary strata represented repeated alternations of freshwater and marine conditions.

The Paris Basin research very much influenced his more general theory, which was put forward as the preface to a substantial publication on fossil bones. This preface was quickly recognised to be of major importance and went through several editions as a *Discours sur les révolutions de la surface du globe.* The Scottish Wernerian geologist Robert Jameson translated this as an *Essay on the theory of the earth* (Cuvier 1817). Cuvier considered himself to be an empiricist who refused to indulge in speculations that could not be firmly backed up by evidence. The following quotation indicates that **actualism**, the study of currently operating processes as a means to interpret the past, was nothing new to him, but that he considered that it could not be validly applied to older rocks. The Ancient was decoupled from the Recent.

> "We now propose to examine those changes which still take place on our globe, investigating the causes which continue to operate on its surface... This portion of the history of the earth

> is so much the more important, as it has long been considered possible to explain the more ancient revolutions on its surface by means of these still existing causes... but we shall presently see that unfortunately this is not the case in physical history; the thread of operations is here broken, the march of Nature is changed, and none of the agents that she now employs were sufficient for the production of her ancient works".

Cuvier argued that the continents had been repeatedly inundated, either by invasions of the sea or by transient floods, after which there had been a general fall of sea level, so that the continents were now in an emergent state. He emphasised that these fluctuations had been by no means gradual. Sudden, dramatic change was particularly evident in the case of mammoths that had been found frozen in Siberian permafrost terrain in the late eighteenth century. Thus:

> "[The last catastrophe] also left in northern countries the bodies of great quadrupeds, encased in ice and preserved with their skin, hair and flesh down to our own times. If they had not been frozen as soon as killed, putrefaction would have decomposed the carcasses. And on the other hand, this continual frost did not previously occupy the places where the animals were seized by the ice, for they could not have existed in such a temperature. The animals were killed, therefore, at the same instant when glacial conditions overwhelmed the countries they inhabited."

This event was only the latest of a succession of catastrophes however. Cuvier went on to say:

> "The dislocations, shiftings, and overturnings of the older strata leave no doubt that sudden and violent causes produced the formations we observe, and similarly the violence of the movements which the seas went through is still attested by the accumulations of debris and of rounded pebbles which in many places lie between solid beds of rock. Life in those times was often disturbed by these frightful events. Numberless living things were victims of such catastrophes: some inhabitants of the dry land, were engulfed in deluges; others, living in the heart of the sea, was left stranded when the ocean floor was suddenly raised up again; and whole races were destroyed forever, leaving only a few relics which the naturalist can scarcely recognise".

The evidence for episodic catastrophes was based on the abrupt disappearance of whole faunas up the stratigraphic succession, for which his personal experience was restricted to the Paris Basin, and alternations of freshwater and marine strata in the same region, and conglomerates and intense tectonic disturbance of strata in older rocks, which has been widely reported since the late eighteenth century. Cuvier showed no published interest in attempting to reconcile the geological record with *Genesis*, let alone to invoke supernatural causes for his catastrophes; the principal focus of attention was to determine the record of past events as precisely as he could. While the successive catastrophes, whose origins were not speculated upon, caused the extinction of animal species, the species that replaced them in the stratal succession were thought possibly to have migrated from other continents or seas. The implication is that not all the catastrophes, though extensive, had necessarily been worldwide. Cuvier did not speculate on how the new species originated but he strongly opposed the evolutionary views of his colleague Lamarck.

Cuvier's most outstanding catastrophist disciple was Léonce Elie de Beaumont, who in 1829 put forward a theory that attempted to integrate knowledge of mountain ranges with the stratigraphic record now being established by the use of fossils in correlation. He determined the age of formations of different mountain ranges from the position in the stratigraphic record of the oldest undisturbed strata flanking them. He followed Cuvier in arguing that folded and tilted strata implied sudden disturbance, and that one was not entitled to extrapolate such "catastrophic" phenomena from the manifestly slow and gradual "causes now in operation". Long periods of quiescence were interrupted suddenly by relatively short-lasting upheavals of the land or torrential inundations of the sea. Twelve episodes of mountain elevation were inferred, which had a drastic effect on contemporary organisms, causing large-scale extinctions. Thus it was clear that the elevation of the Pyrenees had taken place at the end of the Cretaceous, and there were profound differences between Cretaceous and Tertiary faunas. Elie de Beaumont also thought he could determine the age of mountain ranges by their regional orientation or strike. His synthesis, which was much more catastrophist than the Wernerian doctrine it replaced, was enthusiastically received by his geological contemporaries and had a great influence on the European continent for several decades (Greene 1983).

Another Frenchman, the palaeontologist and stratigrapher Alcide d'Orbigny, was also Cuvier's disciple. Armed with the fuller knowledge of the stratigraphic record obtained from several decades of research, he attempted to generalise Cuvier's picture of successive catastrophic extinctions of marine faunas to a global scale, and proposed a succession of stratigraphic stages for the Jurassic and Cretaceous systems, punctuated by extinction events. Though such a generalisation was decidedly premature in the light of contemporary knowledge his basic concept of the stage as a tool for global correlation has survived, and such terms as Cenomanian and Bathonian remain standard today.

Another strand of catastrophist thought was more marked in Britain. In the 1820s many geologists in this country were firm believers in the scriptural Deluge, and Jameson, in his editorial introduction to the translated version of Cuvier's *Discours*, endeavoured to equate Cuvier's most recent revolution with this. Preoccupation with the Deluge was of much more concern in Britain than on the continent, probably because some of the leading figures, such as Buckland, Conybeare and Sedgwick, were clergymen. The leading diluvialist (or student of the Deluge) was William Buckland, of the University of Oxford. According to him the Deluge had been universal, and therefore could not precisely correspond to Cuvier's most recent revolution which had been attributed to a general interchange in position of continents and oceans. He used reports of fossil bones found at high altitudes in the Andes and Himalayas as evidence that the great diluvial event was not confined to low ground but was deep enough to cover the higher mountains. He mustered a wide range of evidence in support of the Deluge. This included: gorges and ravines cut through mountains, buttes and mesas, and immense deposits of gravel, together with scattered boulders on hills or slopes where no river could have carried them. It was impossible to refer such phenomena to the puny modern agents of erosion and sediment transport. Buckland in fact supported Sir James Hall's interpretation of some kind of huge torrent resembling a tidal wave of enormous magnitude.

Diluvialism was in active retreat in the 1830's, and was quickly abandoned by all reputable geologists, including eventually Buckland himself. This was in no small part due to the influence of Scrope's careful geomorphological research in the Massif Central of France. Scrope was able to demonstrate convincingly that the present valleys could, granted only sufficient time, have been eroded by the existing streams. The present valley of the Loire and its tributaries within the Le Puy Basin have been excavated since lava was erupted, sections through which occur on the valley sides. Yet the lavas are undeniably contemporary because cones of loose scoriae occur locally. These would have quickly been removed by any violent rush of water. It follows that the erosive faces of the existing streams must be sufficient to excavate the valleys. Furthermore it was difficult to see how a rushing torrent could have excavated the complex bends in flood-plain rivers known as meanders. Scrope thought that one could extrapolate such results to other mountainous regions where the evidence of volcanic rocks is lacking. The Cambridge geologist Sedgwick, in his 1830 Presidential Address to the Geological Society of London, was readily prepared to concede, with Scrope, that river meanders argue against catastrophism, and referred to other evidence favouring gentle rates of river erosion. But Scrope's ideas were, he thought, conspicuously unsuccessful in accounting either for far-travelled erratic boulders and the gravels of the Thames Valley and elsewhere. Processes operating at the present day seemed too feeble to account for these phenomena. Not until well over a decade later did it become generally accepted that such erratics and gravels signified the presence in quite recent geological times of extensive land ice (Hallam 1983).

Scrope's work made a great impression on Charles Lyell, who produced a three-volume synthesis of geological knowledge (Lyell 1833) which was to prove an eminently successful and influential work, running into no fewer than twelve editions, the last appearing in 1875, the year of his death. The subtitle of his book indicates his purpose, "to explain the former changes of the earth's surface by reference to causes now in operation". He considered that one of the major impediments to the advance of geological knowledge was the limited duration accorded to past time. Most of the first volume of the *Principles of Geology* and all of the second describe geological processes now in action, with the aim of demonstrating that these processes operating at the present intensity are adequate to account for all the geological phenomena of the past, given eons of time. Thus:

> "[These mountains could have been uplifted slowly, by a series of small steps.] We know that during one earthquake the coast of Chile may be raised for a hundred miles to the average height of about three feet. A repetition of two thousand shocks, of equal violence, might produce a mountain chain one hundred miles long, and six thousand feet high. Now, should one or two only of these convulsions happen in a century, it would be consistent with the order of events experienced by the Chilians from the earliest times: but if the whole of them were to occur in the next hundred years, the entire district must be depopulated, scarcely any animal or plants could survive, and the surface would be one confused heap of ruin and desolation".

Catastrophists had pointed to mixtures of terrestrial and marine fossils in certain strata as proof of devasting revolutions of the kind envisaged by Cuvier. Lyell insisted that strata containing such mixtures are forming today in the Mississippi delta. Here the waters of annual floods regularly carry parts of terrestrial plants and bones of terrestrial animals into lagoons or shallow waters offshore, where they are buried with skeletons of marine fish and shellfish. Lyell's doctrine of **uniformitarianism**, as it came to be called, was well received only in part by most of his contemporaries, and modern historians of science have accused him of tendentiousness and distortion in his essentially polemical treatment of the earlier generation of geologists. Gould (1965) has distinguished two separate strands of thought in Lyell's original concept:-

1. Methodological uniformitarianism, comprising the assumptions that natural laws are constant in space and time, and that no hypothetical unknown processes be invoked if observed historical results can be explained by presently observable processes.

2. A substantive uniformitarianism, which postulates a uniformity of material conditions or of rates of processes.

Now clearly what Gould calls methodological uniformitarianism is vital for interpreting the past. There is, however, no logical connection with substantive uniformitarianism, and no good reason to assume uniformity of rates of processes. Thus there may well have been times when rates of continental erosion, say, were greater than now. Biological evolution is decidedly non-uniformitarian and must have effected considerable changes of environment. Geology is a historical science concerned with past configurations of the earth dealing with successions of unique, strictly unrepeatable events through time. In this it differs fundamentally from more theoretical sciences like physics, which deal with the finding and testing of universal laws.

It is in what has been called **directionalism** that the greatest conflict arose between Lyell's views of a steady-state planet and his somewhat misleadingly termed "catastrophist" opponents (Hallam 1983). The consensus of informed opinion in the 1830s was that many geological phenomena, such as mountain building, required the expenditure of greater energy at times in the past, and it happens that Lyell's basic assumption was criticised in a penetrating way by the person who coined the terms catastrophism and uniformitarianism, William Whewell (1837):

> "*Time*, inexhaustible and ever accumulating his efficacy, can undoubtedly do much for the theorist in geology; but *Force*, whose limits we cannot measure, and whose nature we cannot fathom, is also a power never to be slighted: and to call in one, to protect us from the other, is equally presumptuous to whichever of the two our superstition leans. In reality, when we speak of the uniformity of nature, are we not obliged to use the term in a very large sense, in order to make the doctrine at all tenable? It includes catastrophes and convulsions of a very extensive and intense kind; what is the limit to the violence which we must allow to these changes? In order to enable ourselves to represent geological causes as operating with uniform energy through all time, we must measure our time by long cycles, in which repose and violence alternate; how long must we extend this cycle of change, the repetition of which we express by the word *uniformity?* Any way must we suppose that all our experience, geological as well as historical, includes more than *one* such cycle? Why must we insist upon it, that man has been long enough an observer to obtain the *average* of forces which are changing through immeasurable time".

One of Lyell's few uncritical supporters in these early years was the young Charles Darwin, whose own excellent geological work on volcanoes, coral reefs and mountain ranges was greatly influenced by the *Principles.* Lyellian gradualism also had a considerable effect on his biological thought. It is thus a great irony that his theory of evolution, in which speciation took place gradually by means of a succession of numerous small steps, implied on a grand scale a strong directionalism to the history of life, something which troubled Lyell greatly and delayed his conversion for some years. On the subject of catastrophism, however, Darwin's views would have been heartily endorsed from the start. "The old notion of all the inhabitants of the Earth having been swept away by catastrophes at successive periods is very generally given up" (Darwin 1859). He was firmly of the opinion that biotic interactions, such as competition for food and space, were of considerably more importance in promoting evolution and extinction than changes in physical environment. In his widely shared view dramatic faunal turnovers in the stratigraphic record implied *ipso facto* major erosional gaps in the rock succession. Thus the strong contrast between Mesozoic and Kainozoic faunas, recognised by Phillips as early as 1840, did not elicit much interest last century in terms of extinction.

In the latter half of the century uniformitarianism came under strong attack from a quite different source, the eminent Scottish physicist William Thomson, more generally known by his name after elevation to the peerage, Lord Kelvin. Early in his career he had developed an interest in the relationship between the Earth's age and its internal temperature and in his first published work in 1841 had demonstrated his mastery of Fourier's theory of heat conduction, which was to provide him with his key mathematical tool. He had long been convinced that the distribution of heat within the Earth was the key to its age, and indeed the notion that the Earth was originally a hot, molten ball that had cooled gradually dates back to Descartes and Leibnitz. To cut a long story short (Burchfield 1975) Kelvin arrived in 1863 at a best estimate for the age of the Earth of 98 million years. This contrasted markedly with Darwin's estimate of 300 million years for the denudation age of the Weald in south east England, an area with rocks at the surface no older than Cretaceous, based on estimates of erosion rate. Kelvin was scathingly critical of this estimate, and the assumptions on which it was based, and indeed on the uniformitarians' notion of almost unlimited time (Lyell was too canny to risk any even approximate age estimates until very late in his career, under the provocation of Kelvin's work).

Contrary to popular assumptions today, Kelvin's age estimate was well received by many contemporary geologists, and it appeared to give strong support to the directionalist argument, because an older, hotter planet would have experienced tectonic and igneous activity at an appreciable higher rate than today. Only at the end of the century, when Kelvin had become increasingly dogmatic and had pared down his age estimate to a mere 24 million years, did a significant adverse reaction by the geological community set in. The story of how Kelvin's basic assumptions were undermined by the discovery of radioactivity is well known, and it turns out in retrospect that Lyell's assumption of a steady-state Earth is not so far wrong if attention is restricted to the inorganic world of the last 2000 million years (Hallam 1983). This is likely at least to be broadly true of the composition of the lithosphere, hydrosphere and atmosphere (excepting organically-mediated oxygen increase in the Proterozoic and early Palaeozoic), of igneous and tectonic activity, erosion and sedimentation and patterns of climate. Such a broad constancy would have, of course, to encompass long-term episodicities or periodicities in geological phenomena associated with plate tectonics, such as ocean opening and closing, with concomitant formation of mountain ranges, changes of sea level and growth and disappearance of polar ice caps.

The rise of neocatastrophism

Generations of students, at least in the Anglo-American world, have been indoctrinated with uniformitarianism, so that even the dullest of them can recite the ancient litany: the present is the key to the past. In fact this merely expresses actualism, widely practised since the early days of geology by "uniformitarians" and "catastrophists" alike. As we have seen, Lyell's position was more extreme, that there has been an overall constancy through time in geological processes, and that it was unnecessary to invoke catastrophic change in the past to account for any geological or biological phenomena.

Such a view was indeed commonplace among geologists and palaeontologists until quite recently, but something of a reaction set in the nineteen sixties and nineteen seventies with the increasing invocation of relatively uncommon, short-lived episodic events of unusually large magnitude. This can be seen in several fields of research.

(1) *Geomorphology.* The Scablands of eastern Washington State, USA, is an area where deep channels have been eroded, with extensive

scouring, into thick glacial deposits and underlying bedrock; the channel bottoms are filled with coarse gravel brought in from well outside the area. Early this century Bretz (1923) interpreted the channel phenomena as the result of a gigantic, catastrophic flood produced by glacial meltwaters, rather than the slower action of normal stream flow. Unsurprisingly this postulation of an event reminiscent of the Noachian Deluge proved utterly unacceptable to other geologists, but vindication came eventually. This was partly the result of aerial photographs revealing landscape features invisible from the ground. Ridges on the channel flows were revealed which could only plausibly be interpreted as giant ripples, implying a large volume of rapidly flowing water. In addition, geologists working in western Montana discovered evidence of a large glacial lake, a likely source for floodwater. Bretz's catastrophe hypothesis has been fruitful for a much wider region, and Scablands have been found in association with lakes elsewhere in the American West, most notably the immense ancestor of the present Great Salt Lake, Lake Bonneville (Bretz 1969). His ideas have also helped in the interpretation of the famous Martian channels.

(2) *Sedimentology and stratigraphy.* There is now widespread acceptance that sedimentation is frequently a short-lived episodic process interrupted by much longer intervals of non-deposition; this is especially true of siliciclastic sediments in fluvial, lacustrine and marine environments. In many instances nothing may happen for many years until a locally catastrophic storm or tsunami event, and the term tempestite has been introduced for beds with distinctive types of sedimentary structures or shell accumulations thought to signify settling after major storm disturbance (Einsele & Seilacher 1982). These sedimentary phenomena can probably be accommodated comfortably under what has traditionally been understood as uniformitarianism, if averaging out is undertaken over lengths of time appreciably greater than one year. After all, the episodic character of earthquakes and volcanic activity posed no serious problem for Lyell. However, it has also been suggested that such episodic phenomena are characteristic of the stratigraphic record as a whole, which has been compared with the traditional life of a soldier: long periods of boredom interrupted by moments of terror (Ager 1973). In other words the records often indicates long periods of stability, in both

rock and fossil associations, with relatively drastic shifts to a new state, rather than the sort of gradual change a strict uniformitarian might have anticipated.

(3) *Tectonics.* The general acceptance of plate tectonics in the last couple of decades has made geologists more sympathetic to the possibility of significant variations in the past concerning, for example, rates of orogenic and volcanic activity, though such rates of course need not have been "catastrophic" in a global sense. What is much more drastic is the bold proposal that the Earth has expanded by a fifth of its diameter in the last 200 million years, implying a veritable long-term catastrophe in the context of earlier Earth history (Carey 1976, Owen 1976). This strongly iconoclastic view, while welcomed by some biogeographers, has had a highly sceptical reaction from Earth scientists, who have pointed out numerous objections which many would consider insuperable (Hallam 1984c, Weijermars 1986).

(4) *Palaeontology.* The gradualistic Darwinian paradigm for species change was challenged by Eldredge & Gould (1972), who proposed an alternative theory called "punctuated equilibrium", involving long periods of morphological stability and brief episodes of pronounced change. The parallel with neocatastrophist geological thought is obvious, but in no way does the Eldredge and Gould theory seriously challenge the fundamentals of Darwinian evolution, and indeed the authors never claimed that it did. Nor would they consider that the evolution of one species from another involved anything that could reasonably be called catastrophic. To recognise possible catastrophic phenomena on a global scale it is necessary to consider the temporal replacement of whole faunas, a major preoccupation of Cuvier.

Mass extinctions

Mass extinctions are considered to result from the disruption of community structure by a catastrophic event. Catastrophe in this context has been defined by Knoll (1984) as "biospheric perturbations that appear instantaneous when viewed at the level of resolution provided by the geological record." An alternative definition is that of Archibold & Clemens (1982): "a single event that set in motion a chain of other events, thereby causing major biological changes and extinction within at most a few thousand years". It is important to appreciate the limits of time resolution that can be established from the stratigraphic record. In the view of Dingus (1984) it is unlikely that one can distinguish, for the most discussed extinction event across the Cretaceous-Tertiary boundary, episodes of extinction lasting 100 years or less from episodes lasting as long as 100,000 years. In favourable circumstances this upper age limit may be reducible to a few thousand years but doubtfully much less. Sampling problems

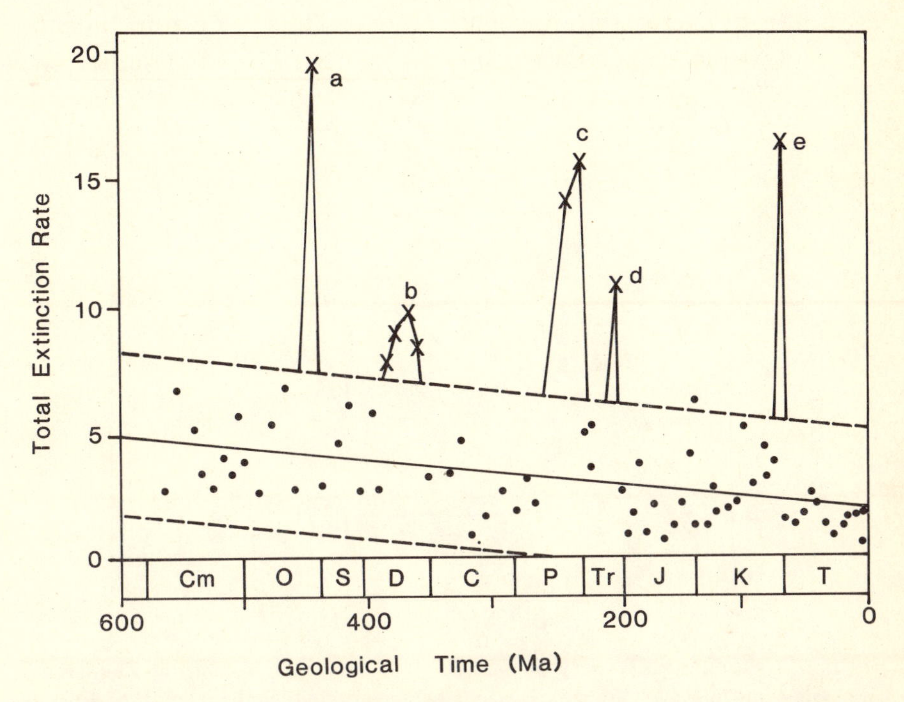

Figure 1. Extinction rate (families/Myr) of marine animals during the Phanerozoic, showing the "big five" mass extinctions as peaks with crosses: (a), late Ordovician; (b), late Devonian; (c), late Permian; (d), late Triassic; (e), late Cretaceous. Background rates (dots) occur within dashed lines; the solid line is a regression. Cm represents Cambrian; O, Ordovician; S, Silurian; D, Devonian; C, Carboniferous; P, Permian; Tr, Triassic; J, Jurassic; K, Cretaceous and T, Tertiary. Adapted from Raup & Sepkoski (1982):

can be considerable, as will be discussed later.

The Phanerozoic record for marine fossils is for the most part much better than for terrestrial fossils, which needs to be borne in mine in any consideration of mass extinction events. We consider in turn the marine and terrestrial realms:

(1) *Marine realm.* Newell (1967) recognised mass extinctions, based on percentages of families that became extinct, at or near the end of the Cambrian, Ordovician, Devonian, Permian, Triassic and Cretaceous. Raup & Sepkoski's (1982) statistical analysis of extinction rate, expressed as families per million years, confirmed the last five as major episodes of mass extinction (Figure 1) and these have been generally accepted since as the "big five"; they also show up as abrupt decreases in diversity (Figure 2). The best documented end-Cretaceous event shows extinction of about 13% of families, 50%

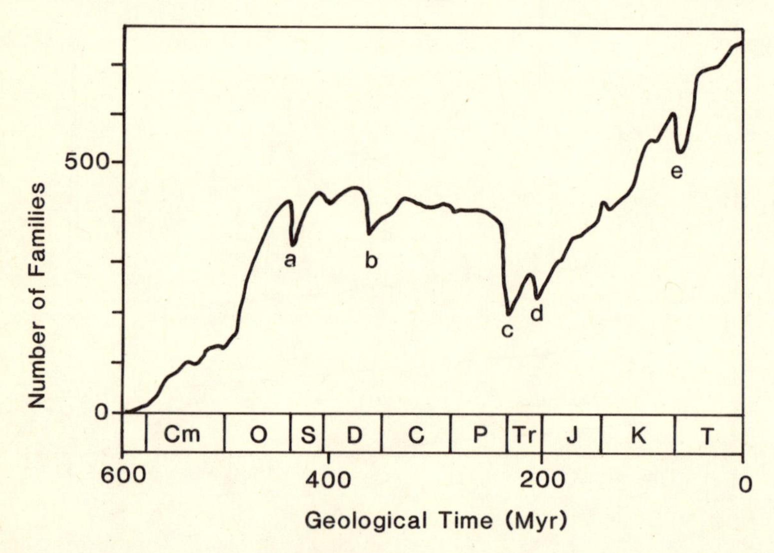

Figure 2. Standing diversity of families of marine animals during Phanerozoic time, with the "big five" mass extinctions being signified by abrupt falls of diversity as follows: (a), -12%; (b), -14%; (c), -52%; (d), -12%; (e), -11%. Abbreviations as in Figure 1. After Raup & Sepkoski (1982):

of genera and 75% of species during the Maastrichtian, the terminal Cretaceous stage, but the biggest of all was at the end of the Permian, when about 50% of shallow-water marine families and an estimated 96% of species became extinct (Raup 1979).

Raup & Sepkoski (1984) subsequently undertook a more detailed study of family extinctions within the last 250 million years, since the late Permian, in which they claimed a regular periodicity of 26 million years for a succession of events (Figure 3). This study, while it has been seized upon with enthusiasm by a number of astronomers, has been subjected to criticism by palaeontologists and statisticians. An objection was quickly raised concerning the reliability of the radiometric timescales utilised (Hallam 1984a) which Raup & Sepkoski countered by repeating their analysis using other timescales and showing that the periodicity persisted although the error margins were greater. Hoffman (1985) argued that Raup & Sepkoski's claims were strongly dependent on a series of arbitrary decisions concerning the

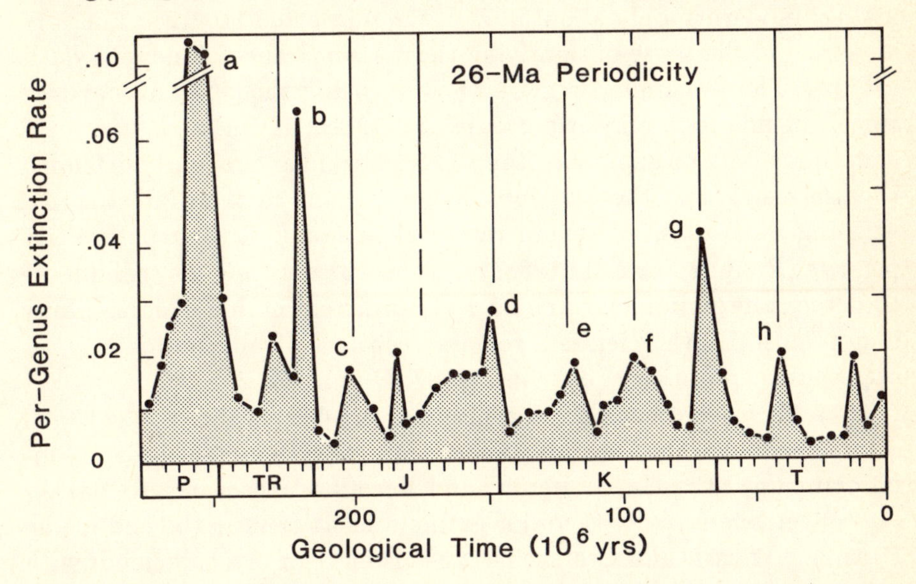

Figure 3. Record of percent extinction (per million years) computed from the record of nearly 1000 genera of marine animals since the Permian, compiled by Sepkoski, which confirms the 26 Myr periodicity (indicated by vertical lines) originally established by Raup & Sepkoski (1984). Ticks along the abscissa signify standard stages. Dots are placed at the centers of the 51 sampling intervals. The signature (a) represents Guadaloupian; (b), upper Norian; (c), Pliensbachian; (d), upper Tithonian; (e), Aptian; (f), Cenomanian; (g), Maastrichtian; (h), upper Eocene and (i), mid Miocene. Adapted from Raup & Sepkoski (1988):

dating of stratigraphic stage boundaries, the culling of the database and definitions of mass extinction events, and that the apparent periodicity could have resulted from purely random processes. Stigler & Wagner (1987) considered that the periodicity could be a statistical artifact. Certain type of measurement error can enhance a periodic signal or cause a pseudoperiodic signal to emerge from aperoidic data. "The hypothesis of a periodic dynamic structure is so powerful in its implications, and so selective in the ease with it imposes itself on us with limited data sets such as this one, that it must be required to pass a stringent test".

Criticism of a quite different sort, namely the quality of the taxonomy utilised, came from Patterson & Smith (1987). With application of the cladistic method of phylogenetic classification to two major groups of which they had specialist knowledge, the fish and echinoderms, only about 25% of the purported family extinctions were real in the sense of signifying the disappearance of monophyletic groups. The remainder is mere "noise", being the result of "extinctions" of non-monophyletic groups, mistaken dating and "families" containing only one species. Raup & Sepkoski have robustly defended themselves against these various criticisms, and in their latest study, utilising a database of genera rather than families, their periodicity persists (Raup & Sepkoski 1988, *cf.* Figure 3) but they are still unable to detect any periodicity among Palaeozoic extinction events. Many if not most Earth scientists remain sceptical of any periodicity, as opposed to episodicity (*e.g.* Lutz 1987, Quinn 1987).

(2) *Terrestrial realm.* Non-marine vertebrates did not get well established until the Devonian and tetrapods such as reptiles and amphibians until the Carboniferous and Permian. According to Bakker (1977) tetrapods suffered major extinctions at or near the end of the Permian, Triassic and Cretaceous, that is more or less coincident with mass extinction episodes in the marine realm. Major extinction events of land mammals in the Kainozoic of North America are reported in the middle Oligocene (Prothero 1985), late Miocene (Webb 1984) and late Pleistocene (Martin 1984). Benton (1985) has carried out a statistical analysis of the non-marine tetrapod record. Although there were several mass extinction events, none was associated with a statistically high extinction rate. The extinction events, including that at the end of the Cretaceous, were apparently the result of a slightly elevated extinction rate. Benton's conclusion, that mass extinctions are statistically indistinguishable from background extinctions, is at variance with that reached for marine fossils by Raup & Sepkoski (1982). Padian & Clemens (1985) stress the inadequate quality of the terrestrial vertebrate record, and its taxonomic treatment, for

analysing phenomena such as mass extinction. For example, the therapsid reptile "flush-crash" in the late Permian may be in part a taxonomic artifact, in part an effect of ecological sampling. Little is known of dinosaur history between the end of the Triassic and the late Cretaceous because of the sparsity of fossils discovered, the only good faunas being in the latest Jurassic of East Africa and the Western United States.

With regard to terrestrial plants, Knoll (1984) and Traverse (1988) agree that the fossil record does not reveal mass extinction events comparable to those recognised in the animal record. It is more accurate to report gradual floral transitions through time, which may be diachronous across the world, and Knoll (1984) accepts a much greater role to displacive competition by adaptively superior organisms, e.g. angiosperms compared with gymnosperms, than for the animal kingdom.

Extinction patterns

Three models are proposed by Kauffman (1986) for the detailed analysis of extinction events: catastrophic, stepwise and graded (Figure 4). Catastrophic extinction would signify the disappearance of a large number of taxa (such as species) in a mere "geological instant", and graded extinction a progressive disappearance over a geologically significant interval of time, though still brief compared with the prolonged periods before and after. A number of practical difficulties have to be faced in testing such models from the stratigraphic record. As Signor & Lipps (1982) have pointed out, the apparent decline of taxa prior to mass extinctions may simply reflect sampling effects and not actual diversity trends. Thus the appearance of a graded extinction can be reasonably generated by sampling effects alone even when the extinction is actually sudden. The sampling problem is minimal for calcareous plankton such as foraminifera and coccoliths because of the huge number of specimens that occur at closely spaced intervals up the stratal succession, and it is these that demonstrate the best evidence for inferring catastrophic change. It is maximal for large, comparatively rare organisms with limited preservation potential, such as dinosaurs. On the other hand, a graded mass extinction may appear locally as catastrophic as a consequence of an erosional hiatus. Disappearance up a given succession therefore does not necessarily signify extinction. The fossil record of a lineage is never 100% complete. There are always time intervals when a taxon is unknown but was evidently present somewhere because it occurs in older and younger time intervals. Such apparent absences produce what has been called a "Lazarus effect" (Jablonski 1986).

Much depends on the taxonomy utilised. Thus a taxonomic "splitter" might recognise stepwise extinctions whereas a "lumper" sees only graded extinction, if each cluster of taxa signifying the steps

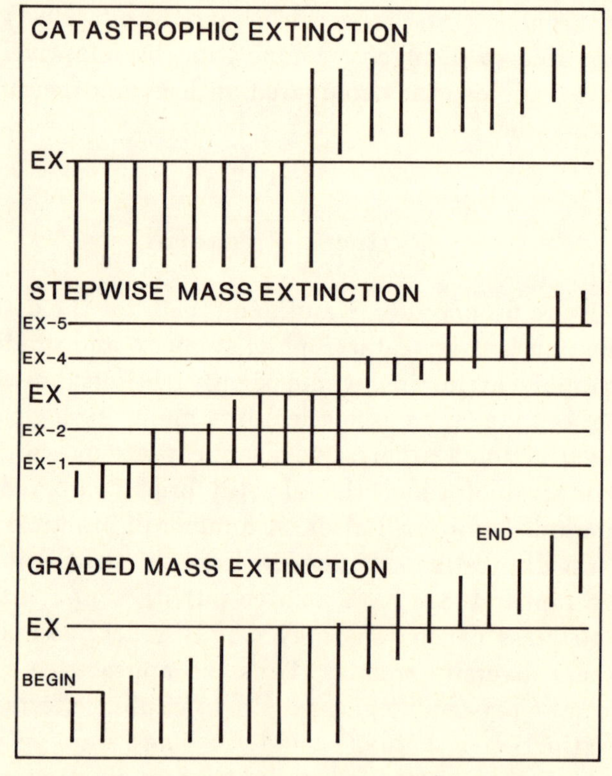

Figure 4. Three models of extinction patterns of taxa, vertical lines representing stratigraphic ranges. Ex-1, Ex-2 etc. define levels of stepwise extinction as components of a single mass extinction event. After Kauffman (1986):

is grouped into one. For this reason, and because of the sampling difficulties mentioned above, it would seem to be extremely difficult to distinguish convincingly stepwise from catastrophic or graded extinction events. With regard to the various mass extinction events that have been recognised, only those across the Eocene–Oligocene and Cretaceous–Tertiary boundaries have been analysed in substantial detail. The deep-sea core record of microfossils across the Eocene–Oligocene boundary, one of Raup & Sepkoski's (1984) examples of periodic extinctions, appears to rule out a catastrophic extinction event. Whether, however, it was a stepwise event, as claimed by Hut *et al.* (1987), or merely graded, has not yet been decisively determined.

For the Cretaceous–Tertiary extinctions, it is generally acknowledged that the planktonic foraminifera and coccoliths underwent a catastrophic extinction, but detailed analysis of stratigraphically expanded sections in Tunisia appears to indicate that the catastrophe extended over several thousand years (Keller 1988, Brinkhuis & Zachariasse 1988). Some benthic invertebrates also underwent catastrophic extinction (*e.g.* Surlyk & Johansen 1984) but a number of important Mesozoic groups that failed to survive into the Tertiary had already been in decline for millions of years, or had already gone extinct, before the end of the Cretaceous. These include ammonites, belemnites and inoceramid and rudist bivalves (Kauffman 1984, Wiedmann 1986). Although the dinosaur record is much poorer, and tolerably good only in a limited area of North America, it also appears to indicate decline in diversity and numbers for up to several million years before the end of the Cretaceous (Sloan *et al.* 1986, Carpenter & Breithaupt 1986). Terrestrial plant extinctions were important only in part of the northern hemisphere, on both sides of the Pacific, and were apparently negligible in the tropics. Short-term ecological trauma apart, extinctions were not catastrophic but extended in time (Johnson & Hickey 1988).

The late Devonian mass extinction event, at the end of the Frasnian stage, has been claimed by McLaren (1983) to be catastrophic, but this is disputed by Farsan (1986) and for the ammonoids is only one of several late Devonian extinction events, according to House (1985). There does, however, appear to have been a spectacularly abrupt disappearance of reef biota at the end of the Frasnian, which is matched by the sudden disappearance of large coral-sponge reefs in the Alps at the end of the Triassic, another of the "big five" (Hallam 1988). Olson & Cornet (1988) claim a correlative abrupt extinction event for reptiles in eastern North America.

Possible causes of mass extinctions

Whereas no doubt a large number of factors have been responsible for extinctions of particular species there are only a few that can plausibly be invoked to account for more or less simultaneous mass extinctions of different types of organism on a global scale. Whereas this can be considered an advantage in attempting to understand causes, it should be noted at the outset that there is an interaction between environmental changes involving climate, volcanicity and sea level, so that determining the critical factor may prove to be not necessarily straightforward. Furthermore, it is important to distinguish between proximate and ultimate causes, particularly in evaluating the relative plausibility of extraterrestrial as opposed to terrestrial controlling factors.

(1) *Bolide impact.* To have any kind of plausibility as a cause for extinction opposed to more conventional explanations based on changes confined to our own planet, a bolide impact scenario must satisfy at least two conditions. The extinctions must be demonstrated by means of the best available stratigraphic evidence to have been on a spectacular scale and to have taken place over geologically negligible periods of time (McLaren 1983). Secondly, there should be independent physical and chemical evidence, such as tektites, shocked quartz and platinum-group metal anomalies. After nearly a decade of intensive research such evidence remains weak to non-existent for all major extinction events except that at the Cretaceous–Tertiary boundary (Orth & Attrep 1988, Orth 1989). Late Eocene deposits contain two regionally extensive tektite horizons but neither correlate with a mass extinction (Glass 1988). Even for the Cretaceous–Tertiary extinction, some would argue that the evidence can be accounted for by a purely terrestrial scenario (Hallam 1987b, McCartney & Loper 1989). Whereas there have undoubtedly been many bolide impacts during the Phanerozoic the only persuasive case for an association with mass extinction is the K-T event, which may therefore be unique in this respect.

Even granted the evidence for impact, it is still by no means clear how many of the end-Cretaceous extinctions relate to it, and no really plausible, exclusively impact-based, "killing scenario" has yet been put forward. Whether involving dust clouds blocking out sunlight, dramatic rise or fall of temperature, acid rain on a massive scale or spectacular wildfires on the continents, they are all too drastic in their environmental effects to account for the selectivity of the extinctions, and the high rate of survival of many groups. The finest time resolution allowable from stratigraphic analysis is rarely less than several thousand years and usually much more. For the end-Cretaceous extinctions it has been argued that, even for the more spectacular events such as those affecting the calcareous plankton, timescales of at least thousand of years must be invoked (Keller 1988, Brinkhuis & Zachariasse 1988). To invoke, in explanation of extinctions over a timespan of up to a million years or more, a series of stepwise extinctions involving multiple impact by comets, as was done by Hut *et al.* (1987), is a significantly weak response to the problem in the absence of independent evidence and requiring as it does acceptance of a dubious astronomical hypothesis (Hallam 1987b, Bailey *et al.* 1987).

(2) *Climate.* Climate can have an indirect influence on sea-level fluctuations at times of substantial polar icecaps as a consequence of glacioeustasy, but a more direct influence has been argued for Phanerozoic extinctions by Stanley (1984, 1986, 1988). His general thesis, that episodes of cooling rather than sea-level falls have promoted mass extinctions in the marine realm, appears to be well supported for the western North Atlantic regime in Plio-Pleistocene times, and perhaps also for earlier in the Tertiary. For pre-Tertiary extinctions, evidence for the climatic cooling hypothesis is decidedly weak, if not non-existent. Thus the biggest extinction event of all, at the end of the Permian, coincides with a time of climatic amelioration following disappearance of the last remants of the Gondwana ice sheet (Veevers & Powell 1987, but see Stanley 1988 for a contrary opinion). Extinctions at the end of the lower Carboniferous and Ordovician could be related to growth of ice sheets, but the key factor in these cases could be the regressions, not the climatic cooling (Brenchley 1984, Powell & Veevers 1987). Stanley's principal argument in favour of his climatic control hypothesis, that organisms with a tropical distribution have been more subject to extinction than

higher latitude organisms, is indecisive because tropical organisms are generally stenotopic and therefore relatively vulnerable to a variety of environmental changes (*cf.* Jablonski 1986).

(3) *Volcanism.* The transmission of large quantities of volcanic dust and aerosols into the atmosphere and stratosphere is known to cause lowering of air temperatures and may also give rise to substantial acid rainfall, with environmentally deleterious consequences (Rampino *et al.* 1988). Volcanism on a massive scale over an extended period of time has been proposed as an alternative to bolide impact to account for the dramatic extinction event at the end of the Cretaceous (Officer *et al.* 1987) but it is unlikely to have been more than a contributory factor to the general extinction scenario (Hallam 1987b).

There is at least one good reason why volcanism is unlikely to have been a major causal factor in Phanerozoic marine extinctions; the correlation between major extinction and volcanic events is rather poor. While the eruption of the Siberian Traps may well coincide quite closely with the end-Permian extinctions (Holser & Magaritz 1987), just as the Deccan Traps eruption apparently coincides with the end-Cretaceous extinctions, the peak of Karoo volcanic activity was early Jurassic not end-Triassic (Fitch & Miller 1984, Aldiss *et al.* 1984) and for the other major extinction events there are no correlative major eruptive events. The Paraná (early Cretaceous), North Atlantic (Palaeocene) and Columbia River flood basalts (Miocene) compare in volume with the Deccan but there are no correlative major extinction events. Where a correlation does exist, as at the end of the Palaeozoic and Mesozoic, both volcanism and sea-level changes may both be expressions of significant events in the mantle (Loper *et al.* 1988).

(4) *Sea-level changes.* A strong correlation exists between both major and extinction episodes in the marine realm and inferred change of sea-level, with the likeliest cause of extinction being bound up with reduction of shallow marine habitat area as a consequence either of regression or the spread of anoxic waters during the subsequent transgression (Hallam 1989; *cf.* Table 1). At times of significant regression the climate on the continents could be expected to exhibit greater seasonal extremes of temperature, a phenomenon that could well have been a major causal factor in the significant increase in extinction rate that took place among land vertebrates at the end of the Permian, Triassic and Cretaceous. It is noteworthy in this

connection that an important mid-Oligocene extinction event among land mammals (Prothero 1985) appears to coincide with a major regressive event.

In seeking an adequate extinction model for Phanerozoic shallow marine invertebrates it is inappropriate to lay too much emphasis on the comparatively abnormal circumstances of the Quaternary, with shallow ocean being represented by pericontinental seas. Rather one should take fully into account the likelihood that the environmental settings of ancient epicontinental seas were in important respects

Extinction events	**regression**	**anoxia**
late Eocene	uncertain	-
end Cretaceous	yes	uncertain
mid Cretaceous (Cenomanian-Turonian)	-	yes
end Jurassic	yes	-
early Jurassic (Toarcian)	-	yes
end Triassic	yes	yes
end Permian	yes	uncertain
mid Carboniferous (Visean-Namurian)	yes	uncertain
end Devonian	yes	yes
late Devonian (Frasnian-Famennian)	yes	yes
end Ordovician	yes	yes
end Cambrian	yes	uncertain
end lower Cambrian	yes	-

Table 1. Correlation of marine regression and anoxic episodes with major Phanerozoic extinction events (Hallam 1989). Extinction events are all included in Sepkoski's (1986) compilation, except that the Toarcian event replaces the Pliensbachian, the immediately older stage.:

different from anything today (Hallam 1981). In such seas, of extreme shallowness over extensive stretches, even a modest change of sea-level could have had significant environmental consequences. Johnson (1974) stressed the importance of organic adaptations to changed circumstances in understanding the likely causes of extinction of neritic invertebrates. During episodes of sustained enlargement of epicontinental seas, organisms become progressively more stenotopic and an equilibrium is established. They are in effect "perched" subject to the continued existence of their environment. Extinctions occur to an extent proportional to the speed of regression, degree of stenotopy attained, or a combination of the two.

Following Johnson (1974) various models relating marine extinctions to regression have been put forward, but none of these models takes into account anoxic events associated with marine transgressions as a possible cause of mass extinction. Such phenomena, however, may in some cases have been more significant than regression. McLaren (1983) has argued that a sudden spread of anoxic bottom waters at the beginning of the Famennian could have been a consequence of oceanic overturn provoked by bolide impact, but the timescale implied by this scenario seems much too short. The widespread black shales at extinction horizons normally have a thickness suggesting deposition over periods of time ranging from at least thousands to hundred of thousands of years. On the other hand the initial spread of anoxic bottom water could have been geologically very rapid, with extinction horizons in the stratigraphic sequence being knife-sharp (e.g. Hallam 1987a). This overcomes an objection raised by McLaren (1983) that regressions were too slow to have caused mass extinctions as dramatically sudden as, for instance, at the Frasnian-Famennian boundary. A combination of regression and subsequent transgression with concomitant spread of anoxic water provides an exceptionally powerful means of substantially increasing extinction rate among epicontinental marine organisms. That not all sea-level falls (e.g. in the early Devonian) correlate with mass extinction episodes, and that there is no obviously simple relationship between the amount of sea-level change and the extent of extinction, probably indicates that the key factor is areal reduction of habitat, and this will depend on such factors as continental configuration and topography.

Towards a theory of terrestrial catastrophism

With regard to the underlying causes of sea-level changes, there remains much uncertainty about their frequency, rate and extent (Hallam 1984b). While a mechanism involving volume change of ocean ridges seems inadequate to account for the frequently short-term events that many stratigraphers recognise, the invocation by Godwin

& Anderson (1985) of numerous glacioeustatic events throughout the Phanerozoic to account for small-scale sedimentary cycles seems implausible in the absence of evidence of substantial polar icecaps for most of this time and the evidence from flora and fauna of extended periods of equability. Although there is now apparently evidence of some high latitude ice during equable periods such as the Jurassic and Cretaceous (Frakes & Francis 1988) this does not establish the likelihood of icecaps large enough to affect sea-level in any significant way. Furthermore, anoxic events would be difficult to explain if such icecaps have persisted through the Phanerozoic (Fischer & Arthur 1977). The most important reason for the association of black shales and transgressive episodes is probably that the early stages of transgression over the continents are characterised by broad stretches of poorly oxygenated shallow water, with restricted circulation with the open ocean, that provide a short transit for organic matter from productive surface water to the bottom sediments. Consequently there is less oxidation and greater retention of organic matter (Hallam 1981, Arthur & Jenkyns 1981). The vertical and lateral extent of anoxic waters during a transgressive event would persumably depend on a variety of factors, such as bottom topography, climate and rate and amount of sea-level rise. Some anoxic (perhaps more safely called hypoxic) events in epicontinental seas might have been generated by upwards expansion of the oceanic oxygen-minimum zone, others by phenomena intrinsic to the seas themselves, and it may yet prove unnecessary to invoke the lack of deep ocean refugia at times of habitat restriction in such seas.

Correlation between mass extinctions and magnetic field reversal patterns at the end of Palaeozoic and Mesozoic, together with changes in sea-level, volcanicity and climate, may be best accounted for by a model involving mantle-core interactions, such as that put forward by Loper *et al.* (1988). Major epeirogenic movements of the continents at these times could have had more spectacular influence on sea level than, for example, changes in seafloor spreading rates. The extend to which some such model could provide an adequate causal mechanism for a greater number of mass extinction events remains uncertain, and clearly requires much further research. Those who have supported bolide impacts as the cause of catastrophic mass extinctions have sometimes argued that the strongest opposition to impact hypotheses has come from geologists and palaeontologists indoctrinated with Lyellian uniformitarianism (*e.g.* Hsü 1986) but it is at least equally likely that they have, for good scientific reasons, not been persuaded by the evidence put forward in support. Indeed there is absolutely no reason why geologists should reject *a priori* the occasional global catastrophe that could severely affect the biosphere, but such catastrophes could well be terrestrial in origin (McCartney & Loper 1989). We remain very ignorant about how the Earth's interior behaves over long periods of time.

References

Ager, D.V. 1973 THE NATURE OF THE STRATIGRAPHIC RECORD. Macmillan, London.

Aldiss, D.T., Benson, J.M. & Rundle, C.C. 1984 *Early Jurassic pillow lavas and palynomorphs in the Karoo of eastern Botswana.* Nature **310**, 302-304.

Archibald, J.D. & Clemens, W.A. 1982 *Late Cretaceous extinctions.* Am. Scient. **70**, 377-385.

Arthur, M.A. & Jenkyns, H.C. 1981 *Phosphorites and paleoceanography.* In OCEANOLOG. ACTA, PROC. 26TH INT. GEOL. CONG., GEOLOGY OF OCEANS SYMPOS., PARIS 1980, 83-96.

Bailey, M.A., Wilkinson, D.A. & Wolfendale, A.W. 1987 *Can episodic comet showers explain the 30-Myr cyclicity in the terrestrial record?* Mon. Not. R. astr. Soc. **227**, 863-885.

Bakker, R.T. 1977 *Cycles of diversity and extinction: a plate tectonic/topographic model.* In PATTERNS OF EVOLUTION AS ILLUSTRATED BY THE FOSSIL RECORD (ed. A. Hallam), 431-478. Elsevier, Amsterdam.

Benton, M.J. 1985 *Mass extinction among non-marine tetrapods.* Nature **316**, 811-814.

Brenchley, P.J. 1984 *Late Ordovician extinctions and their relationship to the Gondwana glaciation.* In FOSSILS AND CLIMATE (ed. P.J. Brenchley), 291-316. Wiley, Chichester.

Bretz, J.H. 1923 *The channeled scablands of the Columbia Plateau.* J.

Geol. **31**, 617-649.

Bretz, J.H. 1969 *The Lake Missoula floods and the channeled scablands.* J. Geol. **77**, 505-543.

Brinkhuis, H. & Zachariasse, W.J. 1988 *Dinoflagellate cysts, sea level change and planktonic foraminifera across the Cretaceous-Tertiary boundary at El Haria, northwest Tunisia.* Mar. Micropaleont. **13**, 153-191.

Burchfield, J.D. 1975 LORD KELVIN AND THE AGE OF THE EARTH. Macmillan, London.

Carey, S.W. 1976. THE EXPANDING EARTH. Elsevier, Amsterdam.

Carpenter, K. & Breithaupt, B. 1986 *Latest Cretaceous occurrence of nodosaurid ankylosaurs (Dinosauria, Ornithischia) in western North America and the gradual extinction of the dinosaurs.* J. vert. Paleant. **6**, 251-257.

Cuvier, G. 1817 ESSAY ON THE THEORY OF THE EARTH: WITH MINERALOGICAL NOTES, AND AN ACCOUNT OF CUVIER'S GEOLOGICAL DISCOVERIES BY PROFESSOR JAMESON. (Reprint of Third Edition: Arno, New York, 1978).

Darwin, C.R. 1859. ON THE ORIGIN OF SPECIES. John Murray, London.

Dingus, L. 1984 *Effects of stratigraphic completeness on interpretations of extinction rates across the Cretaceous-Tertiary boundary.* Paleobiol. **10**, 420-438.

Einsele, G. & Seilacher, A. (eds.) 1982 CYCLIC AND EVENT STRATIFICATION. Springer Verlag, Berlin.

Eldridge, N. & Gould, S.J. 1972 *Punctuated equilibria: an alternative to phyletic gradualism.* In MODELS IN PALEOBIOLOGY (ed. T.J.M. Schopf), 82-115. Freeman, Cooper & Co., San Francisco.

Elie de Beaumont, L. 1829 RECHERCHES SUR QUELQUES-UNES DES RESOLUTIONS DE LA SURFACE DU GLOBE. Crochard, Paris.

Farsan, N.M. 1986 *Frasnian mass extinction – a single catastrophic event or cumulative?* In GLOBAL BIO-EVENTS (ed. O.H. Walliser), 189-198. Springer Verlag, Berlin.

Fischer, A.G. & Arthur, M.A. 1977 *Secular variations in the pelagic realm.* Soc. econ. Paleont. Miner. Spec. Publ. no. **25**, 19-50.

Fitch, F.J. & Miller, J.A. 1984 *Dating Karoo igneous rocks by the conventional K-Ar and $^{40}Ar/^{39}Ar$ age spectrum methods.* Spec. Publ. Geol. Soc. S. Afr. **13**, 247-266.

Frakes, L.A. & Francis, J.E. 1988 *A guide to Phanerozoic cold polar climates from high latitude ice-rafting in the Cretaceous.* Nature **333**, 547-549.

Goodwin, P.W. & Anderson, E.J. 1985 *Punctuated aggradational cycles: a general hypothesis of episodic stratigraphic accumulation.* J. Geol. **93**, 515-533.

Gould, S.J. 1965 *Is Uniformitarianism necessary?* Am. J. Sci. **263**, 223-228.

Greene, M. 1983 GEOLOGY IN THE NINETEENTH CENTURY. Cornell Univ. Press, Ithaca, N.Y.

Hallam, A. 1981 FACIES INTERPRETATION AND THE STRATIGRAPHIC RECORD. W.H. Freeman, Oxford.

Hallam A. 1983 GREAT GEOLOGICAL CONTROVERSIES. Oxford Univ. Press, Oxford.

Hallam A. 1984a *The causes of mass extinctions.* Nature **308**, 686-687.

Hallam A. 1984b *Pre-Quaternary sea-level changes.* Ann. Rev. Earth Planet. Sci. **12**, 205-243.

Hallam A. 1984c *The unlikelihood of an expanding Earth.* Geol. Mag. **121**, 653-655.

Hallam A. 1987a *Radiations and extinctions in relation to environmental change in the marine lower Jurassic of north west Europe.* Paleobiol. **13**, 152-168.

Hallam A. 1987b *End-Cretaceous mass extinction event: argument for terrestrial causation.* Science **238**, 1237-1242.

Hallam A. 1988 *The end-Triassic mass extinction event.* Lunar Planet. Inst. Contrib. No. **673**, 66-67.

Hallam A. 1989 *The case for sea-level change as a dominant causal factor in mass extinction of marine invertebrates.* Phil. Trans. Roy. Soc. B. In press.

Hoffman, A. 1985 *Patterns of family extinction depend on definition and geological timescale.* Nature 315, 659-662.

Holser, W.T. & Magaritz, M. 1987 *Events near the Permian-Triassic boundary.* Modern Geol. **11**, 155-180.

House, M.R. 1985 *Correlation of mid-Palaeozoic ammonoid evolutionary events with global sedimentary perturbations.* Nature **313**, 17-22.

Hsü, K.J. 1986 THE GREAT DYING. Harcourt Brace Jovanovich, New York.

Hut, P. et al. 1987 *Comet showers as a cause of mass extinctions.* Nature **329**, 118-126.

Jablonski, D. 1986 *Causes and consequences of mass extinctions: a comparative approach.* In DYNAMICS OF EXTINCTION (ed. D.K. Elliott), pp 183-229. Wiley, New York.

Johnson, J.G. 1974 *Extinction of perched faunas.* Geology **2**, 479-482.

Johnson, K.R. & Hickey, L.J. 1988 *Patterns of megafloral change across the Cretaceous-Tertiary boundary in the northern Great Plains and Rocky Mountains.* Lunar Planet. Inst. Contr. No. **673**, 87.

Kauffman, E.G. 1984 *The fabric of Cretaceous marine extinctions.* In CATASTROPHES AND EARTH HISTORY (eds. W.A. Berggren & J.A. Van Couvering), pp 151-246. Princeton Univ. Press, Princeton, N.Y.

Kauffman E.G.1986 *High resolution event stratigraphy: regional and global Cretaceous bio-events.* In GLOBAL BIO-EVENTS (ed. O.H. Walliser), pp 279-336. Springer Verlag, Berlin.

Keller, G. 1988 *Extinction, survivorship and evolution of planktic foraminifera across the Cretaceous/Tertiary boundary at El Kef, Tunisia.* Mar. Micropaleant, **13**, 239-263.

Knoll, A.H. 1984 *Patterns of extinction in the fossil record of vascular plants.* In EXTINCTIONS (ed. M.H. Nitecki), pp 21-68. Univ. of Chicago Press, Chicago.

Laudan, R. 1987 FROM MINERALOLOGY TO GEOLOGY: THE FOUNDATIONS OF A SCIENCE, 1650-1830. Univ. of Chicago Press, Chicago.

Loper, D.E., McCartney, K. & Buzyna, G. 1988 *A model of correlated periodicity in magnetic-field reversals, climate, and mass extinctions.* J. Geol. **96**, 1-15.

Lutz, T.M. 1987 *Limitations to the statistical analysis of episodic and periodic models of geological time series.* Geology **15**, 1115-1117.

Lyell, C. 1833 PRINCIPLES OF GEOLOGY. **3** vols. John Murray, London.

McCartney, K. & Loper, D.E. 1989 *Emergence of a rival paradigm to account for the Cretaceous/Tertiary event.* J. geol. Education **37**, 36-48.

McLaren, D.J. 1983 *Bolides and biostratigraphy.* Bull.geol.Soc. Am. **94**, 313-324.

Martin, P.S. 1984 *Prehistoric overkill: the global model,* In QUATERNARY EXTINCTIONS: A PREHISTORIC REVOLUTION. (eds. P.S. Martin & R.G. Klein), pp 354-403, Univ. Arizona Press, Tucson.

Newell, N.D. 1967 *Revolutions in the history of life.* Spec. Pap. Geol. Soc. Am. No **89**, 83-91.

Officer, C.B., Hallam, A., Drake, C.L. & Devine, J.D. 1987 Nature **326**, 143-149.

Olsen, P.E. & Cornet, B. 1988 *The Triassic-Jurassic boundary in eastern North America.* Lunar Planet. Inst. Contrib. No. **673**, 135-136.

Orth, C.J. 1989 *Geochemistry of the bio-event horizons.* In MASS EXTINCTIONS: PROCESSES AND EVIDENCE. (ed. S.K. Donovan.) pp. 37-72, Belhaven, London.

Orth, C.J. & Attrep, M. 1988 *Iridium abundance measurements across the bio-event horizons in the geologic record.* Lunar Planet Inst. Contrib. No. **673**, 139-140.

Owen, H.G. 1976 *Continental displacement and expansion of the Earth during the Mesozoic and Cenozoic.* Phil. Trans, Roy. Soc. A **281**, 223-291.

Padian, K. & Clemens, W.A. 1985 *Terrestrial vertebrate diversity: episodes and insights.* In PHANEROZOIC DIVERSITY PATTERNS (ed. J.W. Valentine), pp 41-96. Princeton Univ. Press, Princeton, N.Y.

Patterson, C. & Smith, A.B. 1987 *Periodicity of extinctions: a taxonomic artefact.* Nature **330**, 248-251.

Powell, C. McA. & Veevers, J.J. 1987 *Namurian uplift in Australia and South America triggered the main Gondwanan glaciation.* Nature **326**, 177-179.

Prothero, D.R. 1985 *North American mamalian diversity and Eocene - Oligocene extinctions. Paleobiol.* **11**, *389-405.*

Quinn, J.F. 1987 *On the statistical detection of cycles in extinctions in the marine fossil record.* Paleobiol. **13**, 465-478.

Rampino, M.R, Self, S. & Strothers, R.B. 1988 *Volcanic winter.* Ann. Rev. Earth Planet Sci. **16**, 73-100.

Raup, D.M. 1979 *Size of the Permo-Triassic bottleneck and its evolutionary implications.* Science **206**, 217-218.

Raup, D.M. & Sepkoski, J.J. 1982 *Mass extinctions in the marine fossil record.* Science **215**, 1501-1503.

Raup, D.M. & Sepkoski J.J. 1984 *Periodicity of extinctions in the geologic past.* Proc. Nat. Acad. Sci. **81**, 801-815.

Raup, D.M. & Sepkoski J.J. 1988 *Testing the periodicity of extinction.* Science **241**, 94-96.

Rudwick, M.J.S. 1972 THE MEANING OF FOSSILS: EPISODES IN THE HISTORY OF PALAEONTOLOGY. Macdonald, London.

Sepkoski, J.J. 1986 *Phanerozoic overview of mass extinction.* In PATTERNS AND PROCESSSES IN THE HISTORY OF LIFE (eds. D.M. Raup & D. Jablonski), pp 259-256. Springer Verlag, Berlin.

Signor, P.W. & Lipps, J.H. 1982 *Sampling bias, gradual extinction patterns and catastrophes in the fossil record.* Geol. Soc. Amer. Spec. **190**, 291-298.

Sloan, R.E., Rigby, J.K., Van Valen, L.M. & Gabriel, D. 1986, *Gradual dinosaur extinction and simultaneous ungulate radiation in the Hall Creek Formation.* Science **232**, 629-633.

Stanley, S.M. 1984 *Marine mass extinction: a dominant role for temperature.* In EXTINCTIONS (ed. M. Nitecki), pp 69-117. Univ. Chicago Press, Chicago Press, Chicago.

Stanley S.M. 1986 EXTINCTION. Scientif. American Library, New York.

Stanley S.M. 1988 *Paleozoic mass extinction: shared patterns suggest global cooling as a common cause.* Am. J. Sci. **288**, 334-352.

Stigler, S.M. & Wagner, M.J. 1987. *A substantial bias in non- parametric tests for periodicity in geophysica data.* Science **238**, 940-942.

Surlyk, F. & Johansen, M.B. 1984 *End-Cretaceous brachiopod extinctions in the chalk of Denmark.* Science **223**, 1174-2277.

Traverse, A. 1988 *Plant evolution dances to a different beat: plant and animal evolutionary mechanisms compared.* Histor. Biol. **1**, 277-302.

Veevers, J.J. & Powell, C. McA. 1987 *Late Paleozoic glacial episodes in Gondwanaland reflected in transgressive - regressive depositional sequences in Euramerica.* Bull. geol. Soc. Am. **98**, 475-487.

Webb, S.D. 1984 *Ten million years of mammal extinctions in North America.* In QUATERNARY EXTINCTIONS: A PREHISTORIC REVOLUTION (eds. P.S. Martin and R.G. Klein), pp 189-210. Univ. Arizona Press, Tucson.

Weijermars, R. 1986. *Slow but not fast global expansion may explain the surface dichotomy of Earth.* Phys. Earth Planet Int. **43**, 67-89.

Whewell, W. 1837 HISTORY OF THE INDUCTIVE SCIENCES. Cass, London.

Wiedmann, J. 1986 *Macroinvertebrates and the Cretaceous-Tertiary boundary.* In GLOBAL BIOEVENT, ed. O.H.Walliser), pp 397-410, Springer-Verlag, Berlin.

HYPERVELOCITY IMPACT CRATERING: A CATASTROPHIC TERRESTRIAL GEOLOGIC PROCESS

Richard A.F. Grieve

Geophysics Division, Geological Survey of Canada, Ottawa, Ontario, Canada

Summary. Hypervelocity impact cratering on the Earth is a natural consequence of the character of the Solar System. Approximately 120 terrestrial hypervelocity impact craters are currently known. The known sample is biased towards younger structures on the stable cratonic areas, with the average cratering rate being $5.4 \times 10^{-15}\,km^{-2}\,yr^{-1}$ for structures with diameters greater than 20 km. Impact craters have two major morphological forms: simple bowl-shaped craters, at smaller diameters, and complex structures with structurally uplifted target material in the center and modified rim area, at larger diameters. Terrestrial craters characteristically have a range of shock metamorphic effects indicative of transient pressures of 50 kilobars to megabars. These include shatter cones, microscopic planar features in tectosilicates, diaplectic solid-state glasses and impact melting. In some cases, impact melt rocks contain siderophile anomalies, which indicate the admixture of projectile material. On the basis of the terrestrial cratering record, the physical and chemical evidence at the K-T boundary is consistent with a major impact event. As hypothesized by Alvarez *et al.* (1980) such an event would release the equivalent of $\sim 10^{23}$ joules of kinetic energy. At this time, the terrestrial

cratering record, however, does not provide any direct evidence on whether this was a single or multiple event or on the killing mechanism(s) for the associated mass extinction. Hypervelocity impact is now recognized as a major catastrophic geologic process, with potentially severe biological effects. Most recently, the cratering record has been used to postulate that the Earth is subjected to periodic cometary showers. The evidence for this, however, is equivocal and such postulates should be regarded as highly speculative.

Introduction

Due largely to the results of space exploration programs, impact cratering is now recognized as an important geologic process in the planetary context. For smaller planetary bodies that have retained portions of their earliest crust, impact was clearly a dominant geologic process in early crustal evolution, leading to extensive topographic, thermal and geologic anomalies. For example, large 1000 km-sized impact basins on the moon are responsible for such features as the second order topography, the major gravity anomalies, and the siting of the major volcanic eruptions and related tectonic features.

The Earth is also subjected to Solar System processes. Compared to the other terrestrial planets, however, the record of impact cratering is poorly preserved on the Earth. This is due to the high level of internal activity and the relatively young age and active nature of the Earth's surface. Geologists have been slow to recognize or, in some cases, even accept terrestrial impact craters as a geologic phenomenon. The evidence from space exploration and the hypothesis that the Cretaceous-Tertiary mass extinction was the result of a major impact event, however, have highlighted and exposed impact as a geologic phenomenon to a larger community.

In this contribution, I summarize current knowledge of the terrestrial cratering record, with emphasis on the nature of the process. Arguments are made that the evidence for a major impact at the Cretaceous-Tertiary boundary is fully consistent with the impact record. Some commentary is also offered on suggestions that, based on the terrestrial cratering record, the Earth is subjected to periodic cometary showers, with attendant immediate deterioration in the condition of the biosphere and changes in the geosphere.

Distribution in Space and Time

Approximately 120 terrestrial hypervelocity impact craters are currently known (Figure 1; Grieve 1987). This number does not include the hundreds of known small impact pits, where relatively small mete-

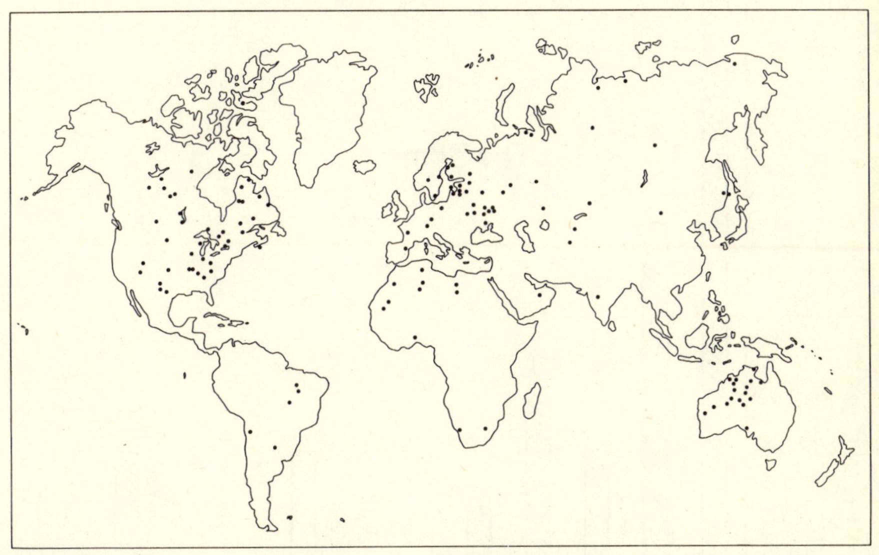

Figure 1. World-wide spatial distribution of known hypervelocity impact craters.:

orites have landed with diminished cosmic velocity due to atmospheric retardation. Hypervelocity impact craters are produced by larger bodies, which impact with undimished or only slightly diminished velocities (Krinov 1963), and are characterized by the production of so-called shock metamorphic effects. Depending on the impact velocity, this is accompanied by the destruction of the impacting body by variable degrees of break-up, melting and vaporization.

Known hypervelocity impact craters range in diameter (D) from approximately 100 m to over 150 km. They range in age from Proterozoic to Recent, with the majority being Phanerozoic in age. Only four are known to be Precambrian and three of these are particularly large: Acraman, D∼160 km, ∼600 Myr old (Williams 1986); Sudbury, D∼190 km, ∼1.85 Byr old (Peredery and Morrison 1984); and Vredefort, D∼150 km, ∼1.97 Byr old (Dietz 1961). The bias towards young ages for known terrestrial craters (Figure 2) reflects problems of crater preservation and recognition. The known record is

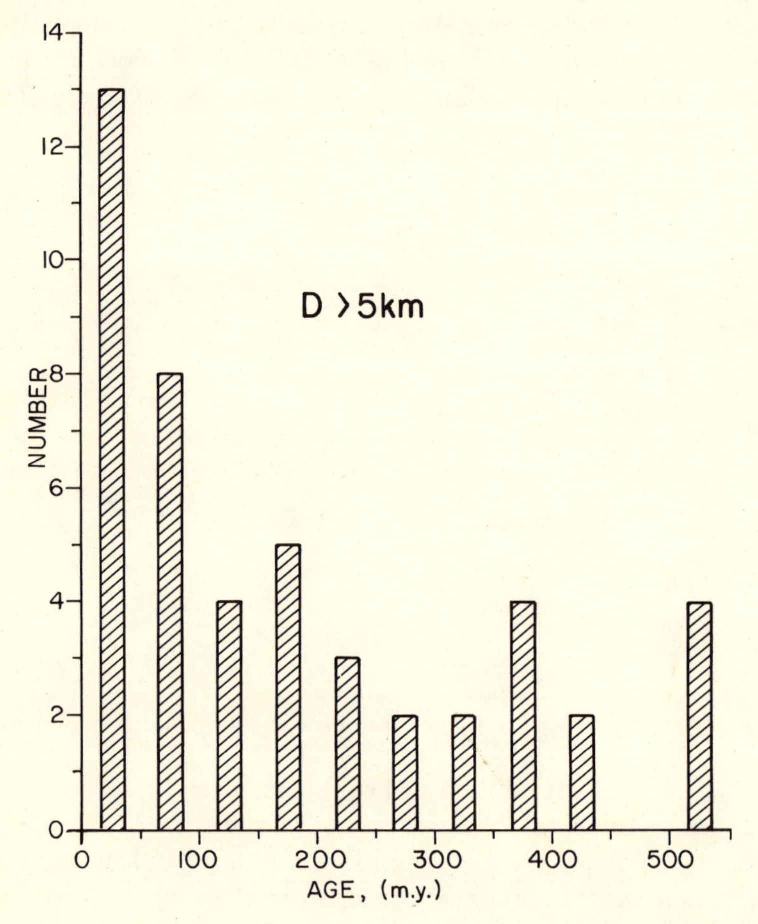

Figure 2. Histogram of ages of known Phanerozoic-aged craters with diameters greater than 5 km. Note bias towards younger ages due to problems of preservation and associated recognition.:

an incomplete sample of a larger population of craters, many of which have been removed by erosion (Figure 3). For example, in recently glaciated areas, it has been estimated that the topographic and subsurface geologic expression of a 20 km-sized crater, not protected from erosion by post-impact sediments, can be unrecognizable in as little as 200 Myr. (Grieve 1984). Many impact craters, however, remain recognizable for longer periods of time. In addition to an age bias, known craters are concentrated in the North American, Australian and European cratons (Figure 1). This reflects the inherent nature of the cratons as relatively stable geologic areas for the acquisition and

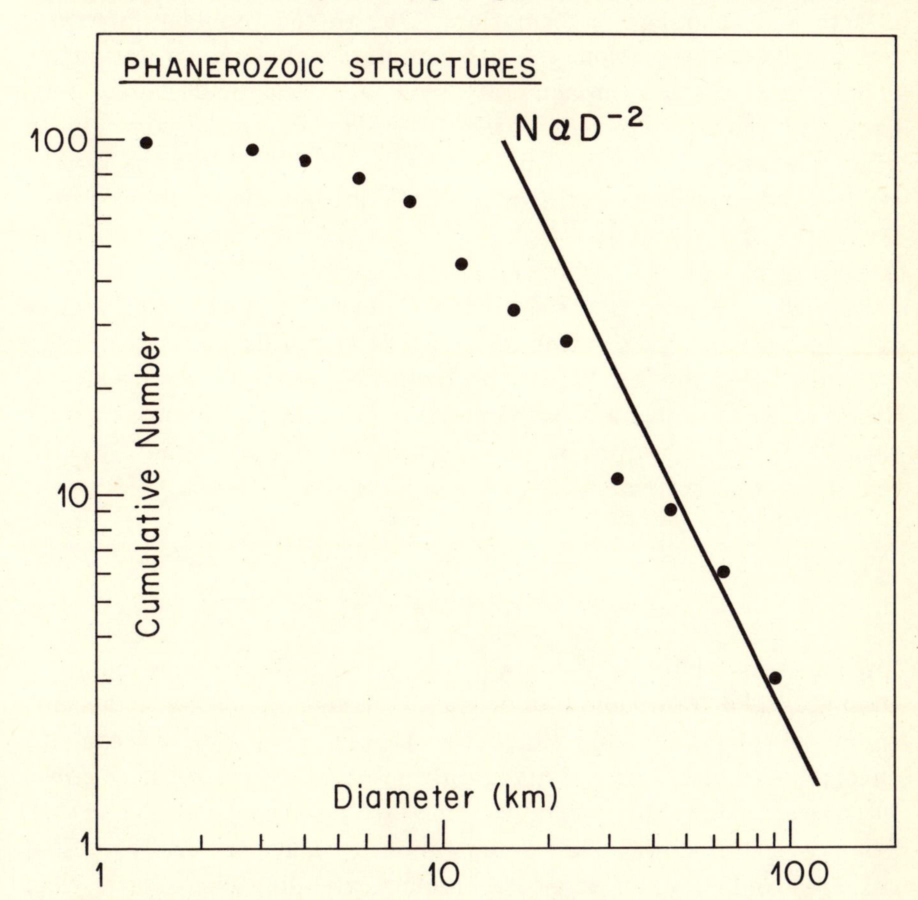

Figure 3. Log - log frequency-size distribution of known Phanerozoic-aged craters with diameters greater than 1 km. Deviation from $N \propto D^{-2}$ distribution of diameters less than ~ 20 km indicates more rapid removal of smaller craters by terrestrial erosion.:

preservation of Phanerozoic-aged craters. These are also areas where there have been active search and research programs in terrestrial impact craters.

The terrestrial cratering rate can be estimated by observation of the size and number of asteroidal and cometary bodies with a probability of collision with the Earth (Shoemaker 1983). According to these studies, the impact of a body sufficient to form a 10 km-sized crater on Earth occurs approximately every 10^5 yr. The cratering rate can also be estimated by direct observation of the numbers of craters and their age of formation. Due to the problems of crater recognition and retention, the size-cumulative frequency distribution of known terrestrial impact craters deviates, at smaller diameters, from the approximately -2 power law (*e.g.* Figure 3) common on other planetary bodies. Thus, only relatively large craters can be used to estimate the cratering rate. The cratering rate estimated from large ($D>20$ km), relatively young (<120 Myr) craters on the North American and Euro-Russian cratons is 5.4×10^{-15} km^{-2} yr^{-1} (Grieve 1984). Using a $N \propto D^{-2}$ size-frequency distribution to approximate the pre-erosional distribution of known craters, this translates to one 10 km-sized crater on Earth approximately every 10^5 yr, which is identical to the estimate based on astronomical observations. Both rate estimates, however, have uncertainties of approximately 50%, reflecting concerns primarily over completeness of search.

Morphology of Impact Craters

Impact craters fall into two major morphological classes: simple and complex. The transition diameter above which simple forms become complex is primarily a function of the size of the event, with secondary effects due to planetary gravity and target strength (Pike 1980). Fresh terrestrial craters are typified by Meteor Crater, Arizona (Figure 4). They consist of a bowl-shaped depression, with a structurally uplifted rim area and an exterior ejecta blanket extending out greater than one crater diameter (Roddy *et al.* 1975). When fresh, the depth to the apparent bottom of the crater is approximately a sixth to a quarter of the rim diameter. Drilling at a number of structures has defined the presence of an interior breccia lens (Figure 5). The breccia is generally clastic, but may have a melted matrix in places, and contain lithic and mineral target rock clasts, with a wide range of shock metamorphic effects (Dence *et al.* 1977). The breccia lens is contained in what is referred to as the true crater (Figure 5). The true crater has a roughly parabolic cross-section and is defined by relatively autochthonous fractured target rocks, which show evidence of shock pressures in the 250 kb range in the center of the true crater floor.

Complex impact structures are characterized by a central uplifted core of shocked target rocks, an annulus of allochthonous breccia or an impact melt sheet and a structurally controlled rim area. They

Figure 4. Aerial view of the 1.2 km diameter Meteor Crater, Arizona, representing a typical fresh simple bowl-shaped crater.:

have relatively shallow depth to diameter ratios, as little as 0.01 – 0.02 (Grieve *et al.* 1981), compared to simple craters and occur at diameters > 2 km in sedimentary and > 4 km in crystalline target rocks on Earth. Morphologic sub-types, for example central peak craters, central peak basins, peak ring basins and multi-ring basins, representing a series with increasing diameter, have been identified on the terrestrial planets (Wood & Head 1976). Similar subtypes occur on the Earth (Figure 6), with the complication that present morphologic features can represent erosional artifacts rather than initial topographic features. In extreme cases, erosion can remove virtually all the impact-related topography, preserving only the roots of the central uplifted core, which can appear as an isolated, eviscerated structural dome, such as Gosses Bluff, Australia; D = 22 km. It is this occasional occurrence of only remanent positive topography, as opposed to a crater *per se*, that has contributed to the reluctance of some workers to accept certain eroded complex structures as having an impact origin.

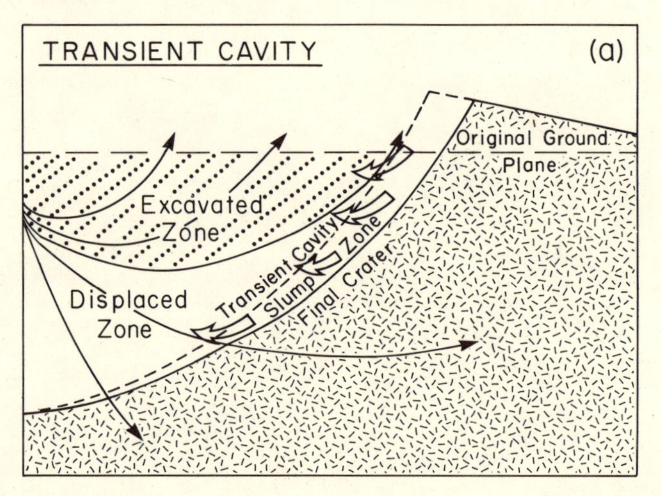

Figure 5(a). **Schematic representation of the formation of the transient crater by the combination of excavation and displacement by the cratering flow-field in the formation of a simple crater. See text for details.:**

Crater Formation and Shock Metamorphism

The r.m.s. impact velocity of bodies in asteroidal-like families with orbits that intersect that of the Earth (Apollos) or that are repeatedly perturbed into Earth-crossing orbits on time-scales of 10^4 yr (Amors) is $\sim$20 km s^{-1} (Shoemaker 1983). Cometary bodies may impact with even higher velocities (Weissman 1982). Thus, considerable kinetic energy is available in the impact of even relatively small bodies.

For example, scaling laws suggest that the 1.2 km diameter Meteor Crater, Arizona, was formed by an iron meteorite only 25–60 m in diameter, which contained $\sim 10^{16}$–10^{17} J of kinetic energy, equivalent to the explosion of ~4–60 megatons of TNT. When a meteoritic body impacts the Earth, the bulk of its kinetic energy is partitioned into kineti~ and internal energy in the target rocks. The exact partitioning of energy is a function of impact conditions but, for average terrestrial impact conditions, it is approximately 80:20 in favour of increasing the internal energy of the target rocks. Increasing the internal energy

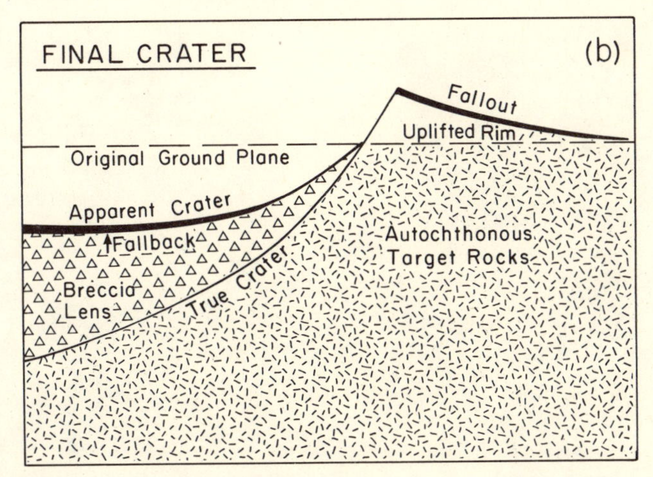

Figure 5(b). Schematic representation of the final form of a simple crater, after collapse of the transient crater walls. Crater is characterized by a structurally uplifted rim and an interior bowl-shaped apparent crater underlain by a breccia lens, the contact of which with relatively autochthonous target rocks defines the true crater. See text for details.:

of the target rocks leads to so-called shock metamorphism, while imparting kinetic energy to the target leads to the ejection of material from the impact site and the formation of a crater.

From small-scale experiments in low-strength material (Gault *et al.* 1968; Stöffler *et al.* 1975), continuum mechanical code calculations (Maxwell 1977; Orphal 1977) and observational data from terrestrial craters (Dence *et al.* 1977; Masaitis *et al.* 1980), the formation of simple craters is fairly well understood. Details and various degrees of quantification of the process can be found in Croft (1980), Melosh (1980) and Grieve & Garvin (1984). In essence, a cavity is formed by the cratering flow field established by accelerations in the target, induced by a hemispherically propagating shock wave and following rarefaction waves. Target material is compressed and accelerated downward and outward by the shock wave, with initial particle velocities of several kilometers per second. Free surfaces, such as the edges of the projectile and the original ground surface, cannot maintain a state of stress and a family of rarefaction or release waves is generated. The rarefaction front trails the shock wave, bringing the compressed and accelerated target material back to ambient pressure. The particle

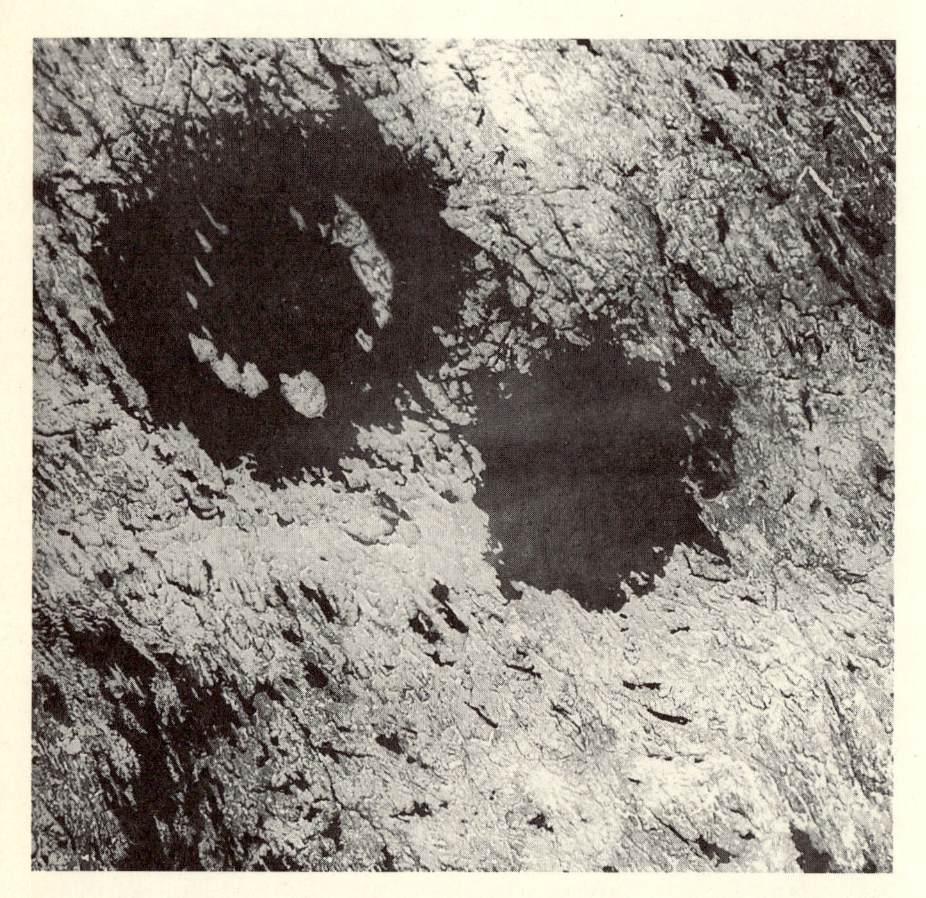

Figure 6. Hand-held Space Shuttle photograph of the twin Clearwater Lake structures. The eastern structure is 22 km in diameter and has a submerged central peak. The western structure is 32 km in diameter and has a ring and submerged central peak.:

decelerations associated with rarefaction lead not only to the slowing of the target material in motion but also to the deflection of the initial radial downward and outward trajectories to upward and outward trajectories for that portion of the target relatively near the surface (Figure 5). Thus, the cratering flow field is established and leads to the growth of a cavity known as the transient cavity. This cavity is formed partly by the upward and outward excavation of material, which travels along ballistic trajectories to form an exterior ejecta blanket, and partly by the downward and outward displacement of material (Figure 5). The transient cavity is highly unstable by virtue of its fractured walls and its depth/diameter ratio of approximately one third (Dence *et al.* 1977), well above the angle of repose. The cavity walls very rapidly collapse inwards, leading to the formation of the interior breccia lens. This breccia lens, therefore, consists of material that never left the cavity. In the freshest examples, it is overlain by a thin layer of true fall-back material, which settled out of the ejecta cloud (Shoemaker 1960).

Uncertainties in scaling experiments and extrapolating computational models to longer times downgrades the constraints available for detailing cratering processes at larger, complex structures. There is, however, clear observational evidence for extensive structural modification of the original transient cavity. In particular, central structures are uplifted from depth, with the amount of uplift approximately one tenth the diameter of the final structure (Grieve *et al.* 1981). Similarly, the final rim is a structural feature, with near-surface lithologies often preserved immediately interior to the rim by virtue of the limitation of excavation to the central area and preservation by downfaulting. The exact details of early-stage transient cavity formation is more debatable. The consensus, however, is that it resembles that of simple crater formation (Schultz & Merrill 1981). In contrast to simple craters, the early-time downward and outward displacements are not locked-in. The cavity collapses as the flow-field in the centre reverses into upward and inward motions due to rebound and gravity collapse of the rim, under essentially hyrodynamic conditions (see Figure 6 in Grieve 1987 for details).

Due to the very transient nature (seconds or less) of compression by the shock wave, as well as local reverberations and variations in the shock compression behaviour of minerals, disequilibrium is common in shock metamorphism compared to endogenic metamorphism. Shock metamorphism also differs in its pressure-temperature regime (Figure 7). Diagnostic shock metamorphic effects are not produced until shock

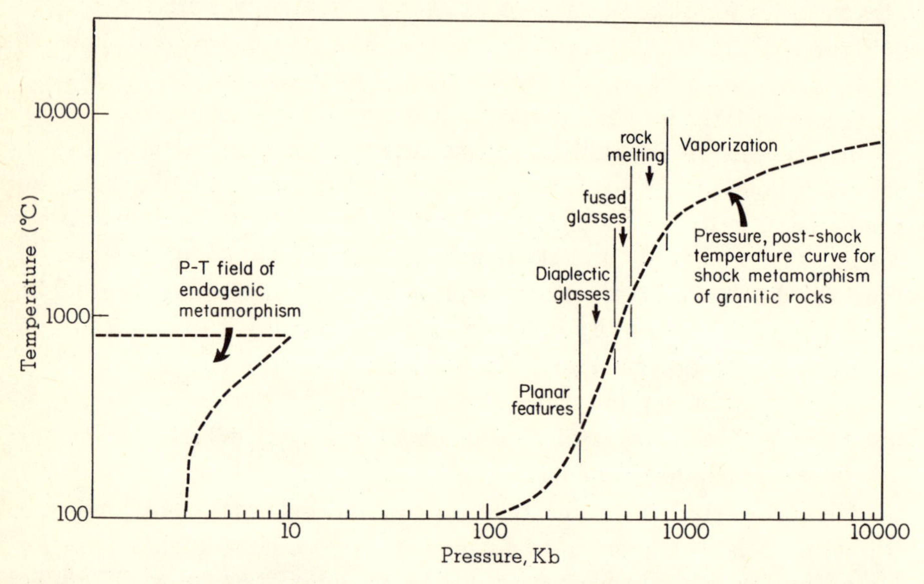

Figure 7. **Log - log pressure-temperature plot of the field of endogenic terrestrial metamorphism and shock metamorphism, with indication of pressures required to form particular shock metamorphic effects.:**

pressures exceed the Hugoniot Elastic Limit of minerals, which is on the order of 50–100 kb, and continue into the megabar range; with accompanying temperatures up to thousands of degrees Celcius.

A considerable literature on shock metamorphism exists, documenting the natural occurrence, the results of laboratory and nuclear explosion experiments and theoretical calculations (*e.g.* French & Short 1968; Stöffler 1972, 1974; Roddy *et al.* 1977). Increasing shock metamorphism represents largely the progressive destruction of short and long range order in minerals. So-called shock lamellae or planar features, representing shear deformation bands filled by glass when fresh, occur particularly in tectosilicates beginning at ~100 kb. The lamellae occur in multiple sets orientated parallel to specific crystallographic planes, with orientation changing and number of sets and their density increasing with increasing shock pressure (Horz 1968; Robertson & Grieve 1977). At higher shock pressures,

~300–400 kb, so-called diaplectic mineral glasses form (Stöffler & Hornemann 1972). At 400–500 kb, mineral fusion glasses form and at higher pressures whole rock melting occurs (Figure 7). Impact melt rocks represent the crystallization product of melted and vaporized target material. As such, they have compositions controlled, not by phase-equilibrium partial melting relations, but by the, often mixed,

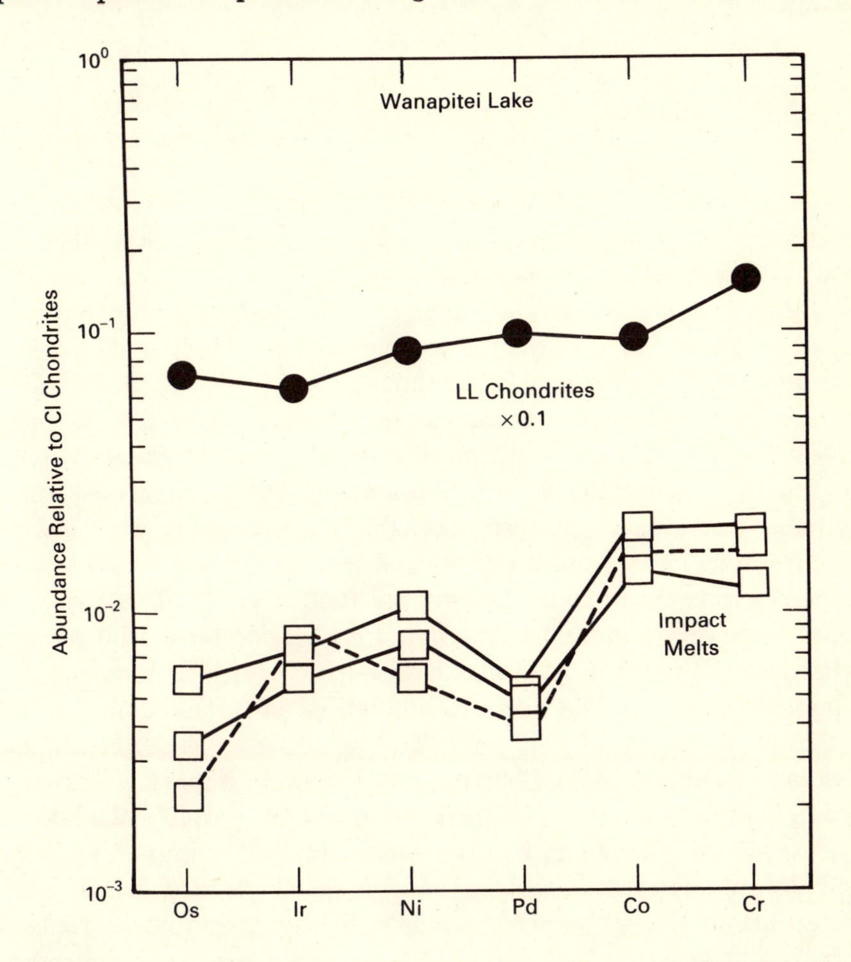

Figure 8. Abundance, relative to CI chondrites, of various siderophile elements and Cr in impact melt rocks from the Wanapitei crater. Siderophiles represent an admixture of < 1% meteoritic material and similar pattern to LL chondrites.:

composition of a volume of the target relatively close to the point of impact (Grieve *et al.* 1977). In some cases, geochemical anomalies in siderophile elements can be attributed to the admixture of projectile material at the general level of less than one percent. From the relative abundance of siderophiles and other elements, such as chromium, it has been possible sometimes to identify the type of impacting body with certainty at specific craters (Figure 8; Palme 1982).

Impact Record and the K-T Event

The suggestion that the Cretaceous-Tertiary (K-T) mass-extinction event was the result of a major impact is not new (*e.g.* de Laubenfels 1956). What has recently elevated this to a working hypothesis is the discovery of chemical and mineralogical evidence in support of a major impact. The original discovery was of elevated Ir and other siderophile elements at a number of K-T boundary sites (Alvarez *et al.* 1980; Ganapathy 1980). Since then, the geochemical anomaly has been confirmed world-wide, in both marine and terrestrial sections. When a number of siderophiles from K-T boundary sites are analysed, their normalised abundance pattern is roughly chondritic (Figure 9), with some fractionation due to possible vaporization and remobilization in the terrestrial environment (Ganapathy 1980; Kyte *et al.* 1985). The situation is very similar to that noted earlier for terrestrial impact melt rocks with traces of admixed meteoritic projectile material (*cf.* Figures 8 and 9). The overall abundances of admixed meteoritic material are variable at K-T sites. They tend, however, to be higher, up to approximately 10% (Alvarez 1986), than in impact melt rocks. This is not a problem. Continuum mechanical code calculations indicate that by far the bulk of the material from the impacting body is accelerated upwards in early-time, high-speed ejecta. It is only such ejecta that are likely to reach stratospheric heights and be distributed globally. In addition, the code calculations indicate that $\sim$ 10–100 times the projectile mass of target material will be accelerated along similar paths to the bulk of the meteoritic material (O'Keefe & Ahrens 1982). Thus, the observed dilution to a few to ten percent meteoritic material in the K-T boundary layer is what is expected.

The physical evidence for impact includes various spherules, high

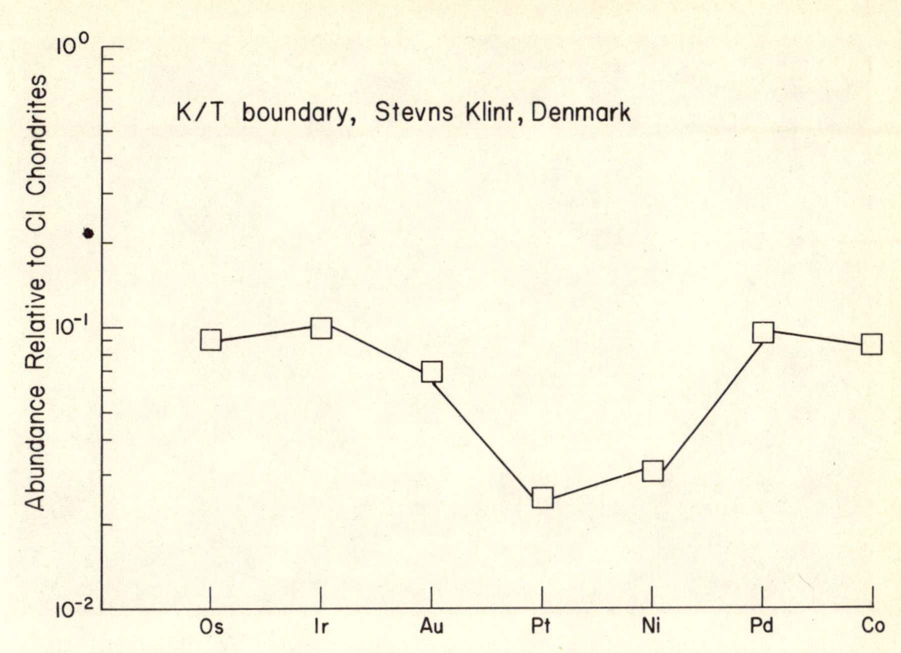

Figure 9. **Abundance, relative to C I chondrites, of various siderophile elements in K-T boundary clay at Stevns Klint, Denmark. Data from Ganapathy (1980). Compare with Figure 8.:**

temperature spinels and shocked clastic grains of quartz and feldspar (Smit & Klaver 1981; Smit & Kyte 1984; Bohor *et al.* 1984). Stishovite, the high pressure polymorph of quartz, has also been discovered in K-T boundary materials (McHone 1988). These products of vaporization, melting and solid-state processes are all known to be produced during the impact process. The most widely quoted physical indicator for impact has been the world-wide presence of planar features (Figure 10) in quartz grains from boundary material (Bohor *et al.*1984). These multiple sets of microscopic planes with specific crystallographic orientations are well-documented from impact craters (French & Short 1986) and nuclear explosions (Short 1968). Recently, Carter *et al.*(1986) have argued that planar features in quartz occur in welded tuffs from the Toba explosive volcanic eruption. However Alexopoulos *et al.* (1988) have observed that, while very rare, lamellar features occur in some Toba samples, they differ in both appearance (indistinct, curved, limited extent, single sets) and orientation from

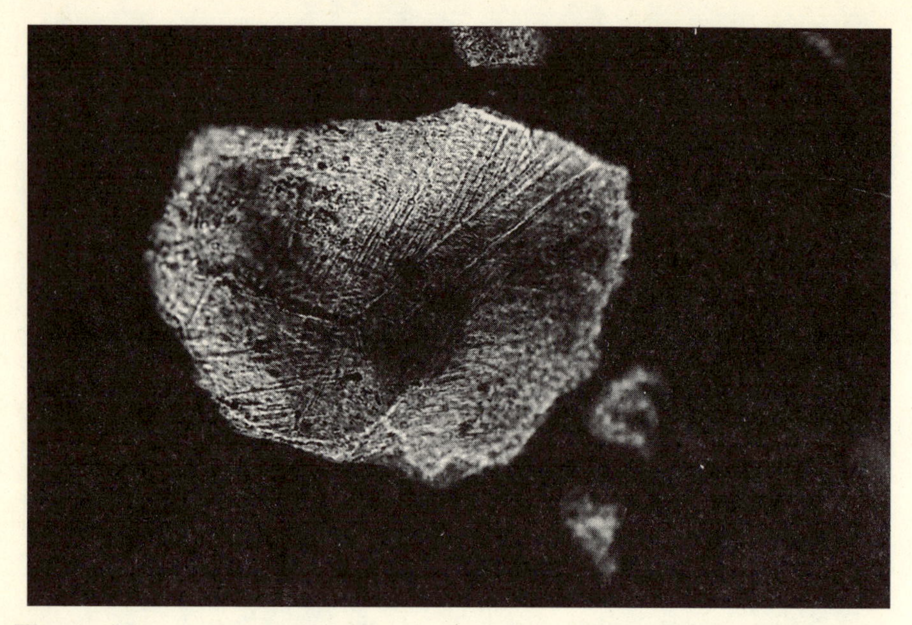

Figure 10. **Photomicrograph of quartz grain showing shock-induced planar features from the Raton Basin. Width of field of view 0.6mm.:**

disgnostic shock produced planar features from impact craters. Thus, terrestrial observations of shock metamorphism are consistent with an impact origin for the planar features in quartz from K-T sites.

The original hypothesis, as given in Alvarez *et al.* (1980), to account for the K-T event calls for the impact of a body of approximately 10^{18}g or $\sim$ 10 km in diameter, depending on its density, to account for the world-wide Ir abundances. The kinetic energy associated with such an event is on the order of 10^{23} J, two orders of magnitude more than the annual release of internal energy (heat, seismic) from the Earth. Although scaling relationships are less certain at these extreme energies, such an event would result in a crater in the 150–200 km size range. On the basis of estimates of terrestrial cratering rate, an event of this size could occur in a 65 million year period.

It has been suggested, on the basis of isotopic chemistry of boundary material, that the K-T event may have been an oceanic impact (DePaolo *et al.* 1983). No impact craters in oceanic crust have been positively associated with the event to date, although the 60 km Montagnais structure has recently been identified on the Nova Scotia shelf (Jansa & PePiper 1987). If a continental impact is involved, as suggested by the presence of shocked quartz and feldspar from a source apparently containing metamorphic, igneous and sedimentary rocks (Owens & Anders 1988), then the K-T crater is unknown at present. This does not represent a fatal flaw in the impact hypothesis. Much of the Earth's land surface is known geologically only at the reconnaissance level. New impact structures are discovered each year, including large and young structures, for example the $>$ 45 km and $<$ 8 Myr old Kara-Kul structure in the Pamirs, U.S.S.R. (Gurov *et al.* 1988).

Some authors have suggested that the K-T event represents a cometary shower (Kyte *et al.* 1985). This could explain the apparent divergence in the isotopic and mineralogical evidence for oceanic and continental impacts. Although some pairs of impact craters are known, for example the East and West Clearwater, Canada and the Kara and Ust-Kara, U.S.S.R., there is no direct evidence in the terrestrial record for cometary showers. To some degree, this suggestion compounds the problem of the undiscovered K-T crater or craters. The only currently known candidate for one such crater is the Manson structure, U.S.A. It has an Ar^{40} age of 66 Myr (Kunk *et al.* 1987) but, at 35 km in diameter, it is too small to be the sole source for the global effects at the K-T boundary.

A number of impact-induced global killing mechanisms have been suggested for the K-T event. These include: darkness with accompanying cold, due to ejecta dust, (Alvarez *et al.* 1980); warming due to greenhouse effects (Emiliani *et al.* 1981); acid rain (Prinn & Fegley 1987); and wildfires, with accompanying soot and darkness (Wolbach *et al.* 1985). Current knowledge of the terrestrial cratering record does not address this question directly. It does, however, suggest that the K-T may have been an exceptional event. There are a number of known impact craters in the $\sim$ 100 km size-range, which have the potential to generate similar effects. For example, the 100 km-sized Popigai structure would have lofted only an order of magnitude less dust than the postulated K-T impact (Grieve & Sharpton, 1986). According to the atmospheric modelling of Toon *et al.* (1982) this would be sufficient to cause the cessation of photosynthesis for $\sim$ 3 months, a period only slightly less than suggested for the K-T. There is, however, no strong evidence linking the Popigai event to a global mass extinction, although its relatively poorly constrained age of 39±9 Myr might be coeval with the Eocene-Oligocene boundary. The apparent lack of extinctions correlated directly with known large impact events suggests that there is some basic threshold size before a major impact event can affect the biosphere. It also suggests that the size of the K-T event may have been underestimated.

Impact Record and Other Events

The linking of a mass extinction with a major impact event has prompted some workers to suggest impact as the causative mechanism for other mass extinctions in the geologic record. This suggestion has been reinforced by claims of a periodicity in the cratering record (Alvarez & Muller 1984; Rampino & Stothers 1984a; and others), which is coincident with a periodicity in the marine extinction record (Raup & Sepkoski 1984, 1986). The entire question of periodicity in the Earth's geologic record is the subject of considerable controversy. The impact record, in particular, has severe problems for time-series analysis, due to the bias in age and size, resulting from erosional effects (*e.g.* Figure 2, 3), and the limitation to relatively small numbers (~20) of large craters (D>5-10 km) with reasonable age estimates. These problems are compounded when the formal uncertainties attached to individual crater age estimates are considered in the analysis (Grieve *et al.* 1987). While more and better constrained ages are required to address these problems, it is likely that questions of periodicity in the cratering record will remain open for a considerable time and may, in fact, require analysis of the more complete lunar, rather than terrestrial, record.

Other claims for periodicities, for example in flood basalt eruptions, tectonism, geomagnetic reversals and so on, have been linked to periodic or episodic impact (Clube & Napier 1982; Rampino & Stothers 1984b; Pal & Creer 1986; and others). There is currently no direct evidence for such suggestions beyond claims of temporal coincidence; however, see Clube, Napier, and Creer & Pal (these proceedings). Although considerable energies are involved in impact, cratering is very much a surface or upper crustal phenomenon. For example, current knowledge of impact cratering mechanics indicate that, although the deepest fracturing occurs in the center of impact structures, most of these fractures are effectively closed by the uplift process (Grieve *et al.* 1981).

Concluding Remarks

Hypervelocity impact on Earth is a consequence of the nature of the Solar System. The record of this process, however, is poorly preserved. Due largely to the importance of cratering as a planetary process, a considerable literature exists on cratering as a geologic phenomenon and current knowledge is sufficient to define to a first-order such aspects as cratering mechanics, cratering rate, energy-scaling and the behaviour of rocks under shock compression. The direct and indirect evidence for a major impact at the K-T boundary is consistent with the known impact record on Earth. The known record does, however, suggest that the K-T impact event may have been underestimated in scope and that it may be unique. Suggestions linking other impact events to biosphere or geosphere changes are, at present, highly speculative. They have, as yet, no direct evidence for a link and should not be confused with or their criticism used to challenge the reality of the K-T event, where there is a wealth of data. The K-T debate has focussed attention on impact as a potentially important geologic process on Earth. It has also highlighted deficiencies in current knowledge, not so much of cratering but of details of ejecta dynamics, atmospheric interactions and energy deposition.

Acknowledgement

This paper is Contribution No 43288 from the Geological Survey of Canada.

References

Alexopoulos, J.S., Grieve, R.A.F. & Robertson, P.B. 1988 *Microscopic lamellar deformation features in quartz: Discriminative characteristics of shock-generated varieties.* Geology, **16**, 796-799.

Alvarez, L.W. Alvarez, W., Asaro, F. & Michel, H.V. 1980 *Extraterrestrial cause for the Cretaceous-Tertiary extinction.* Science, **208**, 1095-1108.

Alvarez, W. 1986 *Toward a theory of impact crises.* EOS, **67**, 649-658.

Alvarez, W. & Muller, R.A. 1984 *Evidence from crater ages for periodic impacts on Earth.* Nature, **308**, 718-720.

Bohor, B.F., Frood, E.E., Modreski, P.J. & Triplehorn, D.M. 1984 *Mineralogic evidence for an impact event at the Cretaceous-Tertiary boundary.* Science, **224**, 867-869.

Carter, N.L., Officer, C.B., Chesner, C.A. & Rose, W.I. 1986 *Dynamic deformation of volcanic ejecta from the Toba caldera: Possible relevance to the Cretacious-Tertiary boundary phenomenon.* Geology, **14**, 380-383.

Clube, S.V.M. & Napier, W.M. 1982 *The role of episodic bombardment in geophysics.* Earth Planet Sci.Lett., **57**, 251-262.

Croft, S.K. 1980 *Cratering flow fields: Implications for the excavation and transient expansion stages of crater formation.* Proc. Lunar Planet. Sci. Conf. 11th, 2347-2378.

de Laubenfels, M.W. 1956 *Dinosaur extinction: One more hypothesis.* J. Paleo., **30**, 207-218.

Dence, M.R., Grieve, R.A.F. & Robertson, P.B. 1977 *Terrestrial impact structures: Principal characteristics and energy considerations.* In IMPACT AND EXPLOSION CRATERING (Roddy, D.J., Repin, R.O. & Merrill, R.B., eds.), 247-276, Pergamon, New York.

DePaolo, D.J., Kyte, F.T., Marshall, B.D., O'Neil, J.R. & Smit, J. 1983 *Rb-Sr, Sm-Nd, K-Ca, O and H isotopic study of Cretaceous-Tertiary boundary sediments, Caravaca, Spain: Evidence for an oceanic impact.* Earth Planet. Sci. Lett., **64**, 356-373.

Dietz, R.S. 1961 *Vredefort ring structure: Meteorite impact scar.* Jour. Geol., **69**, 499-516.

Emiliani, C., Kraus, E.B. & Shoemaker, E.M. 1981 *Sudden death at the end of the Mesozoic.* Earth Planet. Sci. Lett., **55**, 317-334.

French, B.M. & Short, N.M., eds. 1968 SHOCK METAMORPHISM OF MATERIALS. Mono, Baltimore.

Ganapathy, R. 1980 *A major meteorite impact on the Earth 65 million years ago: Evidence from Cretaceous-Tertiary boundary clay.* Science, **109**, 921-923.

Gault, D.E., Quaide, W.L. & Oberbeck, V.R. 1968 *Impact cratering mechanics and structure.* In SHOCK METAMORPHISM OF NATURAL MATERIALS (French, B.M. & Short, N.M., eds.) 87-99, Mono, Baltimore.

Grieve, R.A.F. 1984 *The impact cratering rate in recent time.* Jour. Geophys. Res., **89**, Suppl., B403-B408.

Grieve, R.A.F. 1987 *Terrestrial impact structures.* Ann. Rev. Earth Planet. Sci., **15**, 245-270.

Grieve, R.A.F. & Garvin, J.B. 1984 *A geometric model for excavation and modification at terrestrial simple impact craters.* J. Geophys. Res., **89**, 11561-11572.

Grieve, R.A.F. & Sharpton, V.L. 1986 *The K-T impact event: Some implications from the evidence.* Lunar Planet. Sci. XVII, 289-290.

Grieve, R.A.F., Robertson, P.B. & Dence, M.R. 1981 *Constraints on the formation of ring impact structures based on terrestrial data.* In MULTI-RING BASINS (Shultz, P.H. & Merrill, R.B., eds.), 37-57, Pergamon, New York.

Grieve, R.A.F., Sharpton, V.L., Rupert, J.D. & Goodacre, A.K. 1987 *Detecting a periodic signal in the terrestrial cratering record.* Proc. Lunar Planet. Sci. Conf 18th, 375-382.

Gurov, E.P., Gurova, E.P., Rakitskaya, R.B., Jamichenko, A. Ju. & Monastyretsky, V.I. 1988 *The Kara-Kul lake depression in the Pamirs - a probable astrobleme.* Abst. Eighth Soviet-American Micro-symposium, 35-36.

Horz, F. 1986 *Statistical measurements of deformation structures and refractive indices in experimentally shock loaded quartz.* In SHOCK METAMORPHISM OF NATURAL MATERIALS (French, B.M. & Short, N.M., eds), 243-254, Mono, Baltimore.

Jansa, L.F. & PePiper, G. 1987 *Identification of an underwater extraterrestrial impact crater.* Nature, **327**, 612-614.

Krinov, E.L. 1963 *Meteorite craters on the Earth's surface.* In THE SOLAR SYSTEM, THE MOON, METEORITES AND COMETS (Middlehurst, B.M. & Kuiper, G.P., eds.), 183-207, Univ. Chicago Press, Chicago.

Kunk, M.J., Izett, G.A. & Sutter, J.F. 1987 *$^{40}Ar/^{39}Ar$ age spectra of shocked K-feldspar suggests K-T boundary age for Manson, Iowa, impact structure.* EOS, **68**, 1514.

Kyte, F.T., Smit, J. & Wasson, J.T. 1985 *Siderophile interelement variations in the Cretaceus-Tertiary boundary sediments from Caravaca, Spain.* Earth Planet.Sci.Lett., **73**, 183-195.

Masaitis, V.L., Danilin, A.N., Mashchak, M.S., Raykhlin, A.I., Selivanoskaya, T.V. & Shadenkov, Y.E.M. 1980 THE GEOLOGY OF ASTROBLEMES. Nedra, Leningrad. (In Russian).

Maxwell, D.E. 1977. *Simple Z model of cratering, ejection and overturned flap.* In IMPACT AND EXPLOSION CRATERING (Roddy, D.J., Pepin,

R.O. & Merrill, R.B., eds.), 1003-1008, Pergamon, New York.

McHone, J. 1988 Personal communication.

Melosh, H.J. 1980 *Cratering mechanics-observational, experimental, and theoretical.* Ann. Rev. Earth Planet. Sci., **8**, 65-94.

O'Keefe, J.D. & Ahrens, T.J. 1982 *The interaction of the Cretaceous-Tertiary bolide with the atmosphere, ocean and solid Earth.* Geol. Soc. Amer. Sp. Paper **190**, 103-120.

Orphal, D.L. 1977 *Calculations of explosion cratering - II cratering mechanics and phenomenology.* In IMPACT AND EXPLOSION CRATERING (Roddy, D.J., Pepin, R.O. & Merrill, R.B., eds.), 907-917, Pergamon, New York.

Owens, M.R. & Anders, M.H.. 1988 *Evidence from cathodoluminescence for non-volcanic origin of shocked quartz at the Cretaceous-Tertiary boundary.* Nature, **334**, 145-147.

Pal, P.C. and Creer, K.M. 1986 *Geomagnetic reversals, spurts, and episodes of extraterrestrial catastrophism.* Nature, **320**, 148-150.

Palme, H. 1982 IDENTIFICATION OF PROJECTILES AT LARGE TERRESTRIAL IMPACT CRATERS AND SOME IMPLICATIONS OF IR-RICH CRETACEOUS-TERTIARY BOUNDARY LAYERS. Geol.Soc.Amer.Sp.Paper 190, 223-233.

Peredery, W.V. and Morrison, G.G. 1984 *Discussion of the origin of the Sudbury Structure.* In THE GEOLOGY AND ORE DEPOSITS OF THE SUDBURY STRUCTURE. (Pye, E.G., Naldrett, A.J. and Giblin, P.E., eds.), Ont.Geol.Surv.Sp.Vol. 1, 491-511.

Pike, R.J. 1980 *Formation of complex impact craters: Evidence from Mars and other planets.* Icarus, **43**, 1-19.

Prinn, R.G. and Fegley, B.Jr. 1987 *Bolide impacts, acid rain, and biospheric traumas.* Earth Planets Sci. Lett., **83**, 1-15.

Rampino, M.R. and Stothers, R.B. 1984a *Terrestrial mass extinction, cometary impact and the Sun's motion perpendicular to the galactic plane.* Nature, **308**, 709-712.

Rampino, M.R. and Stothers, R.B. 1984b *Geological rhythms and cometary impacts.* Science, **266**, 1427-1431.

Raup, D.M. and Sepkoski, J.J. 1984 *Periodicity of extinctions in the geologic past.* Proc.Natl.Acad.Sci. U.S.A., **81**, 801-805.

Raup, D.M. and Sepkoski, J.J. Jr. 1986 *Periodic extinctions of families and genera.* Science, **231**, 833-836.

Robertson, P.B. and Grieve, R.A.F. 1977 *Shock attenuation at terrestrial impact structures.* In IMPACT AND EXPLOSION CRATERING. (Roddy, D.J., Pepin, R.O. and Merrill, R.B., eds), 687-702, Pergamon, New York.

Roddy, D.J., Boyce, J.M., Colton, G.W. and Dial. A.L.Jr. 1985 *Meteor Crater, Arizona, rim drilling with thickness, structural uplift, diameter, depth, volume and mass-balance calculations.* Proc.Lunar Sci. Conf. 6th, 2621-2644.

Roddy, D.J., Pepin, R.O. and Merrill, R.B., eds. 1977 IMPACT AND EXPLOSION CRATERING. Pergamon, New York, 1301 pp.

Schultz, P.H. and Merrill, R.B., eds. 1981 MULTI-RING BASINS. Pergamon, New York, 295 pp.

Shoemaker, E.M. 1960 *Penetration of high velocity meteorites, illustrated by Meteor crater, Arizona.* 21st Int.Geol.Congr., Sess. 18, 418-434.

Shoemaker, E.M. 1983 *Asteroid and comet bombardment of the earth.* Ann.Rev.Earth Planet. Sci., **11**, 461-494.

Short, N.M. 1968 *Nuclear-explosion-induced microdeformation of rocks: An aid to the recognition of impact structures.* In SHOCK METAMORPHISM OF NATURAL MATERIALS. (French, B.M. and Short, N.M., eds), 185-210, Mono, Baltimore.

Smit, J. and Klaver, G. 1981 *Sanidine spherules at the Cretaceous-Tertiary boundary indicate a large impact event.* Nature, **292**, 47-49.

Smit, J. and Kyte, F.T. 1984 *Siderophile-rich magnetic spheroids from the Cretaceous-Tertiary boundary in Umbria, Italy.* Nature, **310**, 403-405.

Stöffler, D. 1972 *Deformation and transformation of rock forming minerals by natural and experimental shock processes: I. Behaviour of minerals under shock compression.* Fortschr. Miner, **49**, 50-113.

Stöffler, D. 1974 *Deformation and transformation of rock forming minerals by natural and experimental shock processes: II. Physical properties of shocked minerals.* Fortschr. Miner, **51**, 256-89.

Stöffler, D. and Hornemann, U. 1972 *Quartz and feldspar glass produced by natural and experimental shock.* Meteoritics, **7**, 371-394.

Stöffler, D., Gault, D.E., Wedekind, J. and Polkowski, G. 1975 *Experimental hypervelocity impact into quartz sand: Distribution and shock metamorphism of ejecta.* J.Geophys.Res., **80**, 4062-4077.

Toon, O.B., Pollack, J.B., Ackerman, T.P., Turco, R.P., McKay, C.P. and Liu, M.S. 1982 *Evaluation of an impact-generated dust cloud and its effects on the atmosphere.* Geol.Soc.Amer. Sp.Paper 190, 187-200.

Weissman, P.R. 1982 *Terrestrial impact rates for long and short- period comets.* Geol.Soc.Amer.Sp.Paper 190, 15-24.

Williams, G.E. 1986 *The Acraman impact structure: Source of ejecta in Late PreCambrian Shales, South Australia.* Science **233**, 200-203.

Wolbach, W.S., Lewis, R.S. & Anders, E. 1985 *Cretaceous extinctions: Evidence for wildfires and search for meteoritic material.* Science, **230**, 176-170.

Wood, C.A. & Head, J.W. 1976 *Comparison of impact basins on Mercury, Mars and the Moon.* Proc. Lunar. Sci. Conf. 7th, 3629-3651.

THE CATASTROPHIC ROLE OF GIANT COMETS

S.V.M. Clube

Department of Astrophysics, University of Oxford, Keble Road, Oxford, OX1 3RH, UK

Summary. The currently accepted theories of cosmological and biological evolution both originated at a time when the Earth was thought not to be seriously disturbed by its astronomical environment. The discovery of impact craters and Apollo asteroids has exposed a weakness in such thinking and it has become necessary to examine the whole continuum of catastrophic inputs acting upon the Earth. It is even conceivable now that these inputs and their effects might be incompatible with standard cosmology and Darwinian evolution. At the very least, it is to be expected that the inputs and their effects might place important new constraints on the relationship between cosmology and biology, and there is a good deal of evidence now that new understandings may be achieved in terms of the behaviour of giant comets (diameter $\geq$ 100 km). These bodies still attract very little attention on account of their rarity but they contain the bulk of the cometary mass and arrive in Earth-crossing orbits at $\sim 10^5$yr intervals. There is evidence to suggest their comminution products (including Apollo asteroids, meteoroidal swarms, zodiacal dust) are highly friable and that they interact catastrophically with the Earth on relatively short timescales $\sim 10^2$–10^5yr. On longer timescales $\sim 10^6$–10^{10}yr therefore, the terrestrial response has the character of a discrete

process attributable to successive giant comets whilst the overall flux appears to be episodic and periodic with frequencies of ~ 0.005 and ~ 0.067 Myr^{-1} respectively, due to a possible spiral arm source for giant comets. Such findings place rather severe constraints on the probable nature of the astronomical environment which may extend even to include the cosmological domain if the question of the origin of stars, still unresolved, is bound up with the origin of friable giant comets as well. Whilst the catastrophic rôle of giant comets is thus bound up with very basic astrophysical issues, it is also of interest to identify the remnant of the most recent giant comet in an Earth-crossing orbit since its past *and* future behaviour are likely to be a significant influence on the Earth.

Introduction

Recent searches provide good evidence for a population of ~10^3 asteroids (diameter $\geq$ 1 km) in potential Earth-crossing (Apollo) orbits which are substantially within the region bounded by the main asteroid belt. The planets and their satellites within this region are heavily pock-marked with large craters requiring hypervelocity impacts to account for their formation (Grieve, these proceedings). Within a factor of two or so, these craters correspond in number and size to the frequency of impacts expected for a population of Apollos with the present complement, and it is also likely, on the basis of the cratering record (Figure 1), that the *average* population has been much the same for most of the lifetime of the Solar System. Although the astronomical environment in which geological and biological evolution have taken place is, on this evidence, essentially uniformitarian and harmless, the Earth is bound also to experience global catastrophes due to asteroid encounters at average intervals of ~10^6 yr (Alvarez, these proceedings). Based on the presumed residence time in the stratosphere of dust ejected during the formation of craters, the principal physical effects of such catastrophes may be of only a few years' duration, whilst somewhat longer periods (~ 10^3yr, say) are expected for the processes of natural selection *etc.* that arise as the perturbed biological system undergoes subsequent relaxation.

Whilst these findings have contributed to the recent change in our understanding of the astronomical environment and its effect on the Earth, there are other more general grounds for believing geological and biological evolution are not uniformitarian in the traditional sense. Thus there are recognised degrees of correlation amongst large scale effects in the terrestrial record (*e.g.* climatic and magnetic

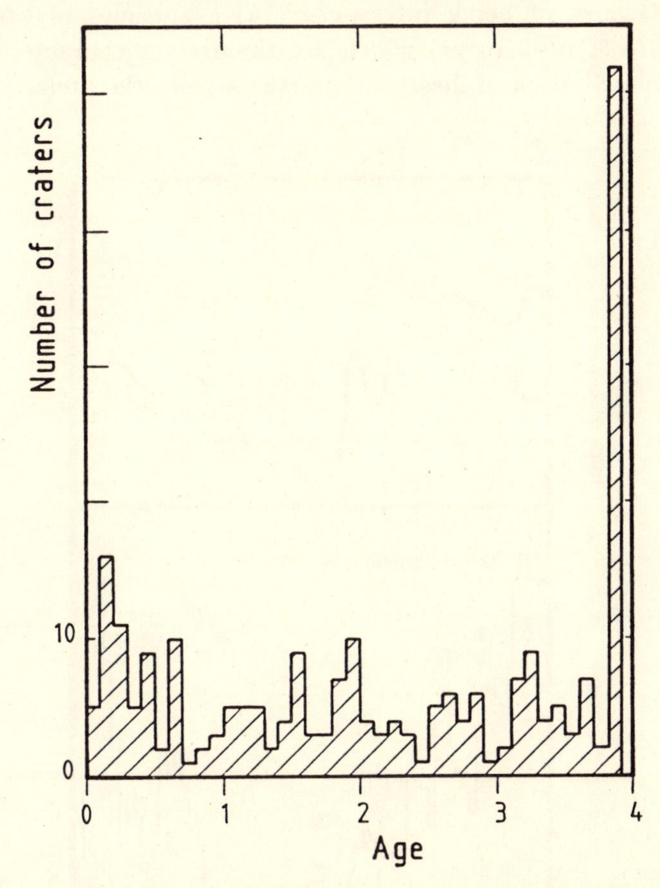

Figure 1. Lunar cratering record for the previous 3.9 Myr (from Baldwin 1985). Note the late 'initial bombardment', the approximately constant average rate thereafter and the seemingly episodic fluctuations (ie variable phase and amplitude) with a putative frequency in the region of 0.005 Myr^{-1}.:

reversals, mantle convection and orogeny, ocean regression, species extinction, speciation *etc.*) such as to suggest recurring activity on the part of a fundamental geophysical process being responsible for the underlying course of evolution. To this extent, the state of the

Earth may likewise be described as uniformitarian. However there is also evidence, based on the same effects, that the fundamental process is *episodic* and *discrete*. If the incidence of geomagnetic reversals (Figure 2), for example, is representative of the evolutionary process as a whole, then an underlying, presumed uniformitarian state has the appearance of being interspersed with episodes of comparable duration ($\sim 3.10^7$–3.10^8yr) which are themselves characterised by a random distribution of discrete departures from the underlying state

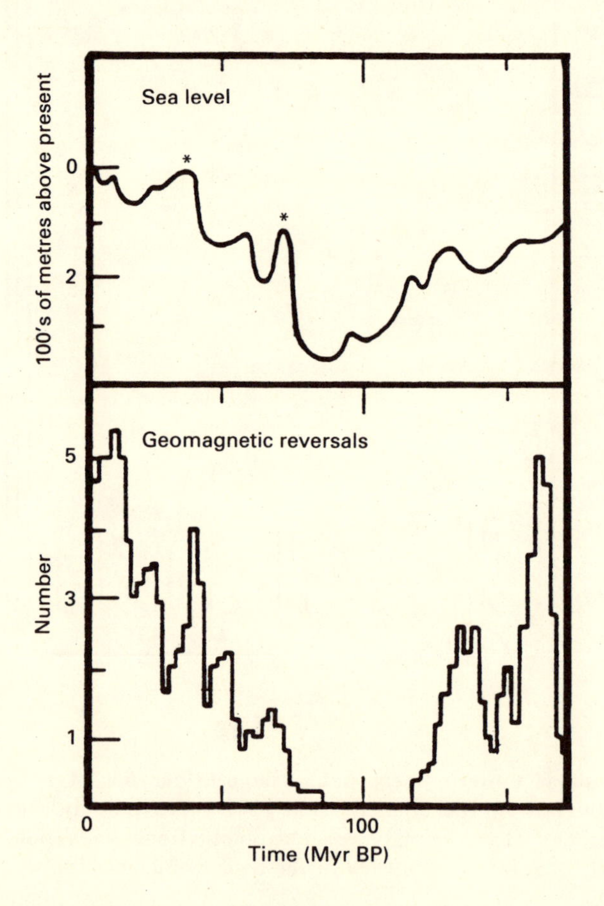

Figure 2. Frequency of geomagnetic field reversals since the end-Jurassic (in Myr^{-1}), compared to the corresponding inverse sea-level variations (based on Hallam 1984). The general correlation is often regarded as suggestive of a physical relationship between climate and mantle convection. Note the single episodic interval in the region of 200 Myr and the periodic frequency $\sim 0.067\ \text{Myr}^{-1}$.:

(apparent duration $\leq 10^5$yr) at intervals of $\sim 10^5$–10^6yr (*i.e.* at an overall frequency not so very different from the average rate of encounters with asteroids).

Fluctuations of this general kind, representing a succession of discrete departures of apparently finite duration from an undisturbed state, run counter to the traditional uniformitarian view of evolution: they are consistent with the general belief now that the overall state of the Earth is neo-catastrophist (Hallam, these proceedings). However there is a variety of conceivable ways in which the terrestrial effects might arise, whilst the rather imperfect calibration and resolution of the geological timescale make for considerable uncertainty in disentangling cause from effect. Thus there is no basis at present, apart from terrestrial chauvinism, for excluding the known astronomical inputs (*e.g.* zodiacal dust, meteoroidal showers, Apollo asteroids) as a possible major cause of these departures from the uniformitarian state. Indeed, on the neo-catastrophist view, it would not be surprising if asteroidal impacts were extreme examples of astronomical inputs giving rise to discrete departures of particular high intensity and brief duration. There is understandable interest therefore in the proposed impact scenario for the KT extinction (Alvarez *et al.* 1980, Smit & Hertogen 1980).

Whilst the extreme astronomical and general terrestrial evidence are compatible with neo-catastrophism, the overall significance and precise mode of action of the various astronomical inputs remain to be determined. Thus it is an important issue whether the scale and frequency of discrete astronomical inputs are such as to dominate or merely influence the course of evolution on very long timescales (10^{10}–10^6yr, say) and whether the microstructure of these inputs on relatively short timescales (10^5–10^2yr, say) is such as to be recognisable in the Pleistocene and Holocene records as well as in the contemporary sky. Indeed the detection in the contemporary sky of *cometary trails* (Sykes *et al.* 1986) and *meteoroidal swarms* (Dorman *et al.* 1978) in Earth-crossing orbits, in addition to the well known *Apollo asteroids*, not to mention the *Tunguska missile* of 1908 (Kresak 1978), now provides rather certain evidence of a regular fragmentation process in the inner Solar System that results in a whole manifold of potentially catastrophic agencies, including *zodiacal dust*, acting on the Earth. On the basis of the observed mass-distribution of comets and the observed friability of the fragmentation products, it is to be generally expected that the latter are chiefly derived from the occasional giant comets (diameter $\geq$ 100 km) that dominate the cometary flux and which arrive in Earth-crossing orbits at intervals $\geq 10^5$yr, thereafter apparently undergoing rapid hierarchical disintegration on timescales $\leq 10^5$yr (see later). The structure and mode of evolution of these bodies, and the nature of the interactions of the various disintegration products with the Earth have naturally become subjects of considerable significance so far as a complete astronomical theory of neo-catastrophism is concerned.

At the same time, whilst the possibility of astronomical inputs is readily understood as perhaps saying something radically new about the fundamental nature of evolution on Earth, there is a tendency to overlook the fact that, until recently, the most prominent inputs, *i.e.* impact explosions, were not expected to occur and were thus far from being perceived as a necessary component of the astronomical environment. It is a completely open question therefore whether the received astronomical framework of the last fifty years, embracing Solar System, Galaxy and Cosmos, in which the observed astronomical inputs are now expected to arise, is in fact compatible with their production. Any proposal, then, in which astronomical inputs are seen as being responsible for a serious change in our perception of how evolution works, *e.g.* involving a change from uniformitarianism to neo-catastrophism, may also have something fundamentally new to say about our general understanding of the Cosmos (Clube 1978, Bailey & Clube 1978, Napier & Clube 1979; *cf.* Appendix).

Episodicity

Whilst the discrete-cum-episodic nature of the terrestrial record may in principle be evidence of a driving force behind evolution that is internal to the Earth, there is no agreement as to the engine involved. Plate motions, for example, which are plausibly associated with mantle convection, appear to be episodic, periods of activity characteristically of $\sim$ 100 Myr ($\pm$ 0.5 dex) being separated by periods of rather similar duration which are comparatively quiescent. It has long been a matter for conjecture however whether mantle convection might be triggered in some way by the possible effects of a lunar torque on the fluid core, interior gravitational settlement or, say, the growth of the "frozen" core at the Earth's centre. Alternatively, mantle convection might simply be one aspect of an essentially intermittent feedback mechanism involving global vulcanism and long-term fluctuations of climate associated with changes in the Earth's moments of inertia which are ultimately unconstrained and which result from relatively sudden displacements of material between the equatorial oceans and polar ice-caps (*cf.* Hallam, *loc. cit.*). Thus, although the discrete nature of the terrestrial record might in principle be due to a fundamental instability in the feedback cycle or the operative engine, the fact that nothing certain has yet been identified, let alone its physics established, means that neither the discrete nor the episodic characteristics are yet understood in purely terrestrial terms.

Despite these uncertainties, episodic fluctuations in the cratering rate appear to be observed (*cf.* Figure 1), bombardments of $\sim$ 100 Myr duration being separated by intervals also of $\sim$ 100 Myr: indeed, on the basis of the correlated meteoroidal flux due to small particles, whose deposition record is unmistakably episodic (Linsday & Srnka 1975), such fluctuations are extremely probable. It seems therefore that the incidence of meteor-induced agglutinates in the lunar regolith is *prima facie* evidence that terrestrial evolution, along with that of other planets, particularly Venus and Mars, may have the episodically variable meteoroidal flux (including Apollo asteroids) as its principal motivating force. An engine of this kind, through its possible perturbations of the Earth's atmosphere, might in fact be the dominant source of climatic change, with derivative inertial displacements at the surface of the Earth modulating mantle convection, which in turn affects both plate tectonics and the fluid motion of the core. According to this model, the pattern of effects influencing evolution is inevitably very complex, involving direct asteroidal impacts, the semi-independent climatic effects and mutagenic properties of meteoroidal showers and zodiacal dust, and such secondary effects as may arise, for example, through continental drift and variations in the geomagnetic dipole.

The overall pattern of evolution based on the *observed* meteoroidal flux, therefore, is one of extended periods of activity (episodes of $\sim$ 100 Myr, say) separated by intervals of similar duration that are relatively undisturbed. Each active period corresponds, it seems, to a random succession of giant comets entering the inner planetary system at $\sim 10^5$–10^6yr intervals and undergoing a hierarchy of disintegrations in short-period orbit which yield dense meteor streams, short-lived swarms of "Tunguska-sized" bodies (100–1000 m) and potential fireballs (1–100 m), many individual asteroids of a kilometre in size or greater, which themselves break up, and at least one conspicuously large core or remnant which is probably enveloped in a swarm of dust and debris (*cf.* Clube & Napier 1984). As already noted, it is to be expected on the basis of the remarkably low density of comets (Rickman 1986) and the extreme friability of much meteoroidal material (Wetherill & Revelle 1982, Fechtig 1982) that the comminution lifetime of such giant comets is in the region of 10^4–10^5yr. During the comminution of one such giant comet, a few dominant impacts are expected but there are numerous lesser ones, the most dramatic taking the form of bolide swarms lasting days or hours (*cf.* Dorman *et al.*, *loc. cit.*). Depending on the cohesive strength of bolide material, *i.e.* the extent to which it is devolatilised during formation and insolation (*e.g.* Bailey *et al.* 1989), such inputs may be responsible for injections into the upper atmosphere of submicron dust particles that essentially remain suspended for periods of a decade or more (introducing a likely association with the greenhouse effect) and/or injections into the lower atmosphere of $\geq$ 50-100 μm dust particles that remain suspended for periods of a year or so (introducing a likely association with climatic coolings). Within the period of evolution of a single giant comet therefore, the incidence of impacts and climatic trauma are largely a matter of chance, depending on the friability of bolides as well as orbital commensurabilities, precession *etc.* of the resultant substreams, though a sustained period of climatic cooling resulting in an ice-epoch (duration $\sim 10^4$yr) is most likely during the earliest stages of giant comet evolution when insolation is the dominant process and a dense meteor stream is more probable than meteoroidal swarms. It is clear nevertheless that the overall effect of giant comets on terrestrial evolution is far more complex than that of single large impacts : in general, so far as mass extinctions are concerned, the climatic trauma will compete with the short-lived destruction due to impacts, whilst there is an expectation of only a loose correlation between impacts – only some of which are due to very friable Apollos – and climate-induced mantle convection with the latter's associated effects of vulcanism and disturbed geomagnetic field.

The meteoroidal flux

There is however no apparent mechanism by which the meteoroidal flux (asteroids, meteoroids, dust) from the asteroid belt may be rendered episodically variable with a characteristic frequency of $\sim$ 0.005 Myr^{-1}. Indeed the known Jovian perturbations are generally considered inadequate so far as the provision of the *current* Apollo asteroid population is concerned (Wetherill 1988). On the other hand, there is an extensive observational basis for associating the meteoroidal flux with Apollo asteroids and both in turn with the cometary flux (*e.g.* Bailey *et al.* 1989; *cf.* Olsson-Steel, these proceedings), and it is for these reasons that astronomers now look mainly to the supply of comets in the inner Solar System to account for possible episodic variations in the general meteoroidal flux.

The supply of comets most commonly associated with the near-ecliptic distribution of Earth-crossing meteoroids and asteroids is the population of short-period comets. The existence of short-period comets in the Solar System is conventionally attributed to the supply of long-period comets from the so-called 'standard Oort cloud'. The transfer from one population to the other is usually regarded as being brought about most efficiently by a sequence of weak Jovian perturbations on a particular set of preferentially oriented orbits in the Oort cloud such as to produce a necessarily flat distribution of short-period comets, as observed (Everhart 1972). Recently this expectation has been challenged (Duncan *et al.* 1988) on the grounds that the true orientation distribution for a properly relaxed population of short-period comets should be isotropic. Thus it has been suggested that there must be a comet belt (analogous to the asteroid belt) far out in the Solar System which supplies most of the short-period comets, albeit through an initial perturber that is not identified. However the conclusion is reached on the basis of dynamical simulations in which the outer planetary masses are significantly increased – for computational speed – thus raising the fundamental question whether the expected secular effects of weak Jovian perturbations are then faithfully preserved. It seems clear therefore that short-period comets are still most plausibly thought of as being derived from the standard Oort cloud though the possibility of a contribution from a necessarily flat cometary belt beyond Uranus cannot at this stage be altogether excluded.

If the Oort cloud is indeed the principal source, as is commonly supposed, the astronomical models that are currently regarded as compatible with possible episodic variations in the general meteoroidal flux may be divided into two broad categories:

(i) Those in which the contributions to the cratering and meteoroidal flux are considered to be more or less equally balanced, in respect of their source, between the asteroid belt and the Oort cometary cloud. The supply from the Oort cloud in this case is considered to be unavoidably concentrated in a succession of intense, short-lived, cometary showers whose individual durations are typically $\sim$ 5 Myr. Proponents of this particular model do not appear to have related the showers in any precise way to episodicity (duration $\sim$ 100 Myr) in either the meteoroidal flux or the terrestrial record, except perhaps to imply that the episodicity must arise through some accidental combination of short-lived showers. Thus it is customary on the basis of this model to take a somewhat limited view of the possible relationship between the meteoroidal flux and terrestrial evolution, simply correlating *mass extinctions* with *cometary showers* and postulating an as yet unseen *inner core* of the Oort cometary cloud to provide the necessarily intense showers (Hills 1981; periodic showers : Alvarez & Muller 1984, Rampino & Stothers 1984; random showers: Hut *et al.* 1987, Bailey *et al.* 1987; *cf.* Bailey, these proceedings).

(ii) Those in which the contributions to the cratering and meteoroidal flux are considered to be dominated by the supply from the Oort cometary cloud, the episodic enhancements in this case reflecting the presence of a standard Oort cloud which is repeatedly captured from star-forming regions during spiral arm passages and then subsequently lost during interarm passages. Both these passages, by reason of the observed solar motion, are of the expected duration ($\sim$ 100 Myr). Proponents of this particular model have no need for a postulated inner core of the Oort cometary cloud (which is *not* in fact observed!); they also argue for a necessarily fundamental relationship between the gross meteoroidal flux (including Apollo asteroids) and most aspects of the terrestrial record (Clube & Napier 1989, Bailey *et al.* 1989; *cf.* Napier, these proceedings).

The existence of a periodic modulation (with harmonics) in the episodic flux, specifically one of $\sim$ 15/30 Myr, is a particularly important question so far as the predictive qualities of these alternative models are concerned. Cyclicity in noisy data is of course by its very nature a contentious issue but the presence of a $\sim$ 15/30 Myr cycle in the terrestrial record has been reported by several investigators and seems to be reasonably secure (Creer, these proceedings; *cf.* Figure 2). So far as category (i) is concerned, the view seems to be generally taken that the 15 Myr component is accidental and that a solar companion star with a period of 30 Myr (which is astrophysically implausible) is required if the proposed inner core of the Oort cloud is to be suitably perturbed. More questionably, it is argued by others that even the 30 Myr cycle is not present in the terrestrial record and that only the random shower model is permissible. The tendency then is merely to associate mass extinctions with cometary showers, to regard the inner core as established fact (quite unjustifiably) and to dissociate astronomy entirely from the episodic character of the meteoroidal flux and the terrestrial record (Hut *et al.* 1987, Bailey *et al.* 1987). So far as category (ii) is concerned, the view is taken that the 15 Myr component is fundamental and that it reflects an astrophysically essential gravitational property of supposedly massive spiral arms (*cf.* Woolley 1965). Thus it is in the nature of massive spiral arms and the smooth galactic disc, acting in concert, to produce oppositely directed local galactic tides which result in Oort cloud comets being relatively unperturbed as the Sun enters and leaves both the upper and lower faces of spiral arms during its vertical oscillations through the galactic plane, thereby substantially reducing the flux of comets in the inner Solar System at the expected intervals of $\sim$ 15 Myr.

Whether the discrete-cum-episodic character of the terrestrial record, reflecting the assumed role of giant comets and spiral arms, is further modulated by an expected galactic periodicity is evidently a crucial question; additional studies *e.g.* of the background iridium flux, vulcanism, sea-level variations *etc.*, are certainly necessary to establish which of these general categories, if either, is more likely to be correct.

Newton's influence

Whilst it might seem rather unsatisfactory that the astronomical and geophysical evidence is broadly compatible with two such extreme models of the Oort cloud (*i.e.* with or without an inner core; one that is permanent or temporary), causing astronomical inputs to be considered either selectively important or fundamental, as the case may be, it is important to realise that this state of affairs is not so much a reflection of permanent uncertainty in science or any basic weakness in the catastrophist viewpoint but more a reflection of the caution that is now necessary in respect of a long-established basic assumption about our environment. Thus there is no doubt that the uniformitarian view has been associated with a general long-standing assumption that astronomical events are largely immaterial so far as the Earth is concerned; and it is also clear that this viewpoint still seriously influences most aspects of science and human behaviour. The emergence of the neo-catastrophist view in recent years, reinforced by the detection of Apollo asteroids and impact craters, not to mention meteoroid swarms and cometary trails, has however initiated a fundamental re-appraisal of our astronomical environment extending from cosmology at one of the scale to the precise nature of the Holocene and the currently observed sky at the other. As we undertake this re-appraisal, therefore, and seek to quantify the interactions that must necessarily occur, the only justifiable approach is to hold *all* the conceivable astronomical options compatible with existing data in readiness for any critical new observational data that may arise, through which one or more such options may then be eliminated. In addition, it is necessary to keep a balanced perspective; in particular, to recognize the original circumstances in which the received view of the Cosmos and Evolution arose.

Even at the time when uniformitarianism was established around 1860, as a contemporary cartoon vividly shows (*vide* the front cover of these proceedings), there were no illusions as to the possible alternative circumstances under which evolution occurred. Indeed it is rather obvious that uniformitarianism could not have arisen had not astronomers of the time been persuaded that cosmic influences were unimportant (*cf.* Davies, these proceedings). As it happens however, astronomers had been persuaded to quite the contrary not long before as a result of the discovery of main belt asteroids and of the observation of a significant meteorite fall in a populated region of northern France. But in the event, an encounter without mishap between the Earth and a swarm of meteors in the Leonid stream was sufficient to reverse the prevailing view. Furthermore Comet Biela, whose orbit at its nearest point to the Earth's orbit was considerably closer than the Moon, was observed to break up in interplanetary space without any harm befalling the Earth. Since there was at this time no clear distinction between comets and meteorites, the evidence of the central decades of the nineteenth century seemed to indicate that comets always disintegrated in interplanetary space or in the atmosphere so as to produce meteor dust or pebbles that were quite harmless. Indeed the conclusion appeared to be all the more secure for the fact that it confirmed important theoretical ideas put forward over a century before by Newton.

To appreciate the significance of Newton's influence in these matters, it is necessary to recall that the period immediately preceding the publication of *Principia*, was a turning point in the study of comets. Thus it was during this period that it came to be realised that comets entered the planetary system on elliptical or near parabolic orbits. Such behaviour was evidently in accordance with the expected behaviour of intrinsically massive bodies in a gravitational field; and it seemed therefore that the inferred property of cometary mass combined with the particular motion of comets must present something of a physical hazard to life on Earth. This viewpoint was very strongly endorsed by Halley while Newton was led to introduce a specific non-catastrophist view of the world in which comets were important, not so much on account of the potential disaster they represented, but on account of the benign influence they represented through the deposition of fresh material on the Earth, arising from their disintegration in space or in the atmosphere. Given what is now known of his unpublished work and of his considerable respect for millenarianism, harking back to an ancient Sumerian-cum-Babylonian astronomical tradition (Keynes 1946), there was a degree of dissembly in Newton's viewpoint. However the need for dissembly is now understood (Clube & Napier 1986; Bailey *et al.* 1989) for Newton and his famous contemporaries, Halley and Whiston, were confronted by an established theological dogma according to which supposed "first causes" did not operate in Nature (*cf.* Thomas 1971). That comets were among the category of proposed "first causes" was a commonplace of the time and it evidently did the reputation of scientists no good at all to be seen harbouring thoughts of very direct action on the part of comets which might also be mistaken for an expression of a seemingly very immediate divine will (*cf.* Figure 3).

Subsequently, Newton's public utterances on this subject were given generous support by the scientific, philosophical and theological communities; and there can be little doubt now that the otherwise rather disturbing meteoric events of the early to mid-nineteenth century were eventually seen as some kind of confirmation of the master's great thoughts. Certainly from the mid-nineteenth century on, along with the establishment of uniformitarianism, the broad sweep of Newton's public cosmological view, which excluded any cosmic influence apart from gravity affecting the Earth, was very widely accepted and science fell into the divisions that are still commonly preserved, in which it is considered to be an affront to commonsense and sound principles to suppose anything drastic can cross the boundaries between astronomy, geology and biology (*cf.* Toulmin 1985).

The supposed insignificance, at the present time, of astronomical inputs in relation to geological and biological evolution, thus constitutes an unjustified extension of an old theological dogma of the counter-Reformation which gained credibility in a more scientific context during the early nineteenth century before any of the physical attributes of comets and meteorites had been properly distinguished. Not only does this supposed insignificance now preserve artificial boundaries between scientific disciplines, it positively inhibits an open-minded approach to the astronomical environment and its potential evolutionary effects. Indeed, with the cratering record and the Earth-crossing asteroid population now revealed, it seems that an appropriate new world-view, embracing the effects of the full range of "small bodies" in the Solar System ($0.1\ \mu m \leq$ diameter ≤ 300 km, say), has become one of the outstanding imperatives of our time.

Figure 3. Sixteenth century fresco depicting the nature of the Final Judgement. Note the realistic portrayal of a meteor shower, its supposed agency. Note also the presumed association with seismic activity typical of Tunguska-sized fireballs. The supposed achievements of pre- and post-Enlightenment enquiry respectively have been the dissociation of meteor showers from the illustrated "first cause" and from their catastrophic influence. (Copyright: Photo Archives – Editions Arthaud).:

The general role of 'small bodies'

It is a common misconception at the present time, arising from particular developments of the subject during the last 10 years, that the sum total of effects on Earth due to the small bodies of the Solar System need amount to no more than an additional source of extinctions at $\sim 10^6$ yr intervals superimposed on an otherwise exclusively terrestrial process (*cf.* category(i) above). This particular catastrophist view now takes much of its strength, paradoxically, from the emphasis given to the particular mass extinction at the KT boundary (Alvarez *et al.* 1980). Indeed it is customary not to mention the mutagenic effects of possible astronomical inputs though these may evidently be as significant as radiative adaptations in determining the future course of evolution (*cf.* Anders, these proceedings). The misconception evidently arises through too simplistic a view being taken of the astronomical environment and a failure to take into consideration the possible effects of small bodies *less than a kilometre in size.* The inevitable result of omitting meteoroidal swarms and zodiacal dust from consideration is a persistent extreme caricature of astronomical catastrophism such that it bears only a partial resemblance to the patently neo-catastrophist character of the terrestrial record. Thus, it is unnecessarily restrictive to think purely in terms of catastrophic extinctions when developing an astronomical theory: as Newton was the first to appreciate, the arrival of cosmic material need not be catastrophic on every occasion. And as we have seen, it is to be expected on the basis of giant comet evolution (*cf.* category (ii) above) that terrestrial evolution is also affected to a significant degree on timescales $\leq 10^6$ yr. In principle, then, the effect of bodies less than a kilometre across, including that of dust, may be towards influencing the extinction/speciation of life on a more frequent or regular basis. This is essentially the viewpoint put forward by Hoyle & Wickramasinghe (1985, chiefly in respect of speciation) and by Clube & Napier (1984, chiefly in respect of the underlying catastrophism).

The zodiacal dust complex contains an estimated $\sim 10^{22}$g of matter in the form of ~ 100 μm grains, and is growing at a net rate of $\sim 10^7$g per second due to the destruction of unidentified meteoroids of mass $\geq 10^2$g in the inner Solar System (*e.g.* Grün *et al.* 1986).

This rapid growth compared to the short dissipation time of zodiacal dust (due to Poynting-Robertson drag, fragmentation and radiative effects) indicates the presence of a copious source which is apparently not specifically related to the general population of readily observed comets and asteroids in the inner Solar System. On the other hand, a substantial fraction of the sporadic meteor flux, presumed to be associated with the zodiacal dust complex, is apparently concentrated in an "eccentric doughnut" centred on the Taurid-Arietid stream (Stohl 1983), whilst the bulk of the fireball flux, associated with larger meteoroids, appears to be in sub-Jovian orbits with aphelia rather similar to Taurid orbits (Dohnanyi 1978). In addition, a rather exceptional number of cometary and asteroidal bodies, *inter alia*, has now been detected in the Taurid-Arietid complex (Table 1) suggesting that this particular system, deriving from a single giant body that arrived in a potential Earth-crossing orbit a few times 10^4 years ago, may be the primary source of the current zodiacal cloud (Clube 1987). If so, it is clearly pertinent to enquire whether the core-body of this huge stream is still present and to examine, if necessary, its likely interactions with the Earth. Thus, a clear demonstration of the existence and of the past and future behaviour of the latest giant comet in Earth-crossing orbit can hardly but reinforce the general significance of these bodies so far as the long-term nature of terrestrial evolution is concerned.

An appreciation of the nature of the interactions with giant comets may be derived from satellite measurements of dust particles in the near-Earth environment (Fechtig, *loc. cit.*). These indicate a significant fraction ($\sim$ 20%) during a year of micron-sized 'zodiacal dust' particles arriving in meteoroidal aggregates of mass $\leq 10^6$g which disintegrate high above the surface of the Earth and the ionosphere, otherwise unnoticed, supposedly due to the acquisition during arrival of an electric charge. Integration over a larger collecting area so as to include larger members of the supposedly top-heavy population of meteoroids suggests the overall input may even be dominated by this influx of micron-sized particles. Thus these observations, along with the estimated shortfall in radar meteor detections (*cf.* Olsson-Steel, *loc. cit.*), point to a regular input of relatively unablated micron-sized particles which descend unobserved through the atmosphere over comparatively long timescales (a few decades, say, depending on size)

Table 1. Probable debris from the most recent giant short-period comets, taken from Bailey *et al.* (1989). The source code is as follows: (1) Whipple & Hamid (1952); (2) Dorman *et al.* (1978); (3) Kresak (1978); (4) Brecher (1984); (5) See text; (6) Sykes *et al.* (1986).:

Object	a (AU)	e	i (deg)	ϖ (deg)	Notes
Meteor streams					
S Taurids	1.93	0.806	5.2	153.2	
N Taurids	2.59	0.861	2.4	162.3	
β Taurids	2.2	0.85	6	162.4	
ξ Perseids	1.6	0.79	0	137	
S Piscids	2.33	0.82	2	104	
N Piscids	2.06	0.80	3	130	
S χ Orionids	2.18	0.78	7	180	
N χ Orionids	2.22	0.79	2	179	
Active comets					
Encke	2.2	0.85	11.9	160	
Rudnicki	—	1.00	9.1	154.7	
Asteroids					
2201 Oljato	2.2	0.71	2.5	172	
1982 TA	2.2	0.76	11.8	128	
1984 KB	2.2	0.76	4.6	146	
5025 P-L	4.20	0.895	6.2	145.8	
2212 Hephaistos	2.1	0.83	11.9	258	
1987 SB	2.16	0.650	2.9	167.4	
Unseen companion	2.4	0.86	—	160	(1)
Impactors on the earth or moon					
Boulder flux					(2)
Boulder swarm					(2)
Tunguska object					(3)
Bruno object					(4)
Larger complexes					
Štohl stream					(5)
Zodiacal cloud					(5)
β Taurid 'trail'					(6)

and which are in principle distinguishable from the relatively more robust interplanetary dust particles, typically of size $\sim$ 50 microns ($\pm$ 0.5 dex), also of meteoroidal origin, which are routinely collected from the stratosphere (Brownlee 1985). Brownlee particles, unlike the smaller particles which are protected within larger aggregates prior to arrival, are known from the presence of solar flare tracks and rare gas implantations, to have survived as independent bodies outside the Earth's atmosphere for periods of time that are consistent with their predicted sojourns in the zodiacal dust complex.

Whilst such micron-sized particles, or less, which are generated in interplanetary space by erosion and fragmentation, are expected to be removed rather rapidly by the effects of the solar wind and radiation (*vide* comet tails), the existence of very fragile, presumably highly devolatilised, meteoroidal aggregates raises the possibility of short-lived, very large clouds of micron-sized particles forming as a result of insolation in the vicinity of any large, disintegrating core-body. Such a large cloud was in fact first detected in the Taurid stream during the latter half of 1974 (Singer & Stanley 1980), and then again, somewhat later in 1975, as a huge meteoroidal swarm on its outward path whilst crossing the Earth's orbit (Dorman *et al.*, *loc. cit.*). The existence of this swarm is plausibly associated with the even larger meteoroid of 1908 giving rise to the Tunguska event (Kresak, *loc. cit.*) and it is a striking coincidence that the orbit of the implied single system is consistent with the subsequent detection by IRAS of a conspicuous dust trail in an orbit very close to that of Comet Encke (Sykes *et al.*, *loc. cit.*).

The balance of evidence suggests we may, in principle, have located the remnant of the most recent giant comet and that it is hidden within a gravitationally bound cloud of debris and dust, an outermost fraction of which is regularly released by the solar tide into the inner Solar System during perihelion passage, thus replenishing the zodiacal complex. Indeed, such a model provides an indication of how giant comets affect the Earth (*cf.* Clube & Asher 1989). Thus, the global deposition of cosmic dust in the decade and a half following the Tunguska event (Ganapathy 1983), much in excess of that expected for the characteristics of the actual explosion (Clube & Napier 1984), suggests that the very friable meteoroidal swarm was likely also to have been involved. Moreover, we cannot exclude the possibility in the aftermath of the 1975 encounter of a significant cosmic contribution to the halogen-rich clouds responsible for the depletion of stratospheric O_3. Any corresponding input in the form of micron-sized particles of low atomic weight (CHON) material associated with a cometary source has the potential, as well, to perturb the atmospheric greenhouse and introduce isotopic anomalies (*e.g.* ^{2}H, ^{10}Be, ^{14}C), leading to an underlying correlation of variations in these anomalies and climate, as observed. On a longer timescale, both effects experience a modulation whose periodic variation must be close to a simple harmonic of the interval of 67 years, apparently consistent with the observed variations of $\sim$200/400yr (Sonett 1990, Raisbeck *et al.* 1990; *cf.* Link 1958).

The study of these physical interactions is evidently still in its infancy but even now, they bear witness, along with the more isolated Tunguska event this century, to the potential catastrophic influence of giant comets on terrestrial evolution, albeit at a very low, current level of intensity. In particular, it is likely now that the latest dead giant comet that gave rise to the Taurid-Arietid complex suffered a major encounter in the asteroid belt some 5000 yr ago (Whipple & Hamid 1952), deflecting the core into a similar but inclined orbit with which that of Comet Encke's is now closely associated. It is clear enough in principle, therefore, that the subsequent near coincidences of the Earth with this core and its attendant trail, especially during the periods of revived activity following a significant subfragmentation, are incidents of some consequence, possibly involving multiple-Tunguska encounters with armageddon-like effects (Clube 1989). The "coincidences" in this instance are limited however to a minimum distance from the core of the trail since the ascending and decending nodes of its orbit at this point in time are close to aphelion and perihelion respectively. On the other hand, the situation was entirely different two and a half millenia ago, each node respectively then coinciding with the Earth's orbit roughly 400 years apart, at times closely corresponding to the beginning and close of Greek civilization. Such knowledge evidently provides an important new perspective on the final centuries of the pre-Christian era when Babylonian astronomy experienced a late revival (Oates 1979), meteoric activity in China seems to have reached some kind of peak (Tian-Shan 1977) and there were generally raised expectations amongst the population at large of impending world-end (Butterfield

1981). The same perspective places in a similar light the preceding, seventh-century Egyptian knowledge, reaching us through Plato's *Timaeus*, of "... a deviation of the bodies that revolve in heaven around the Earth and [of] a destruction, occurring at long intervals, of things on the Earth by a great conflagration". Indeed, our subsequent experience of the Tunguska event provides considerable substance for the warning, also apparently based on experience, that "... once more, after the usual period of years, the torrents of heaven [will sweep] down like a pestilence, leaving only the rude and unlettered among you". Low level catastrophism on 'biblical' timescales, the subject of interest to Newton and his contemporaries (*cf.* Figure 3) as well as to the early nineteenth century catastrophists, is thus a natural extension of the neo-catastrophism derived from the terrestrial record. It deserves rather more attention from twentieth century scientists than it has so far received.

New perspectives

The conclusion that emerges therefore is that giant comets have significance not only for the geological record and evolution as a whole but for the historical record as well. The average Tunguska rate inferred from lunar craters is one every ~500 yr (Shoemaker 1983). However the sense of security provided by this calculation may well be false : for, if long-term episodic-cum-periodic effects are present and disintegrating giant comets are involved, then there are factors allowing for the immediate presence of spiral arms (~ a few), giant meteor streams (~ 10) and meteoroidal swarms (~ 10), giving an average Tunguska rate of nearer one per year during active periods. It follows that for the duration of significant meteoroidal swarms that exist for a century of two every one or two thousand years, we expect ~ 100 Tunguska events (explosive energy ~ 5000 Mt), capable of generating a 'cosmic winter'.

It is known from Chinese records, for example, that there was a significant enhancement of the fireball flux in the Taurid stream during the eleventh century (Astapovic & Terenteva 1968) which is unlikely to have gone unnoticed elsewhere. In fact Pope Urban II is known (Riley-Smith 1986) to have instigated the First Crusade or Holy War in November 1095 at the so-called Council of Clermont in France in the wake of meteor showers that apparently generated widespread eschatological expectations and popular views of the Final Judgement that evidently did not differ from the realistic post-Reformation version (*cf.* Figure 3). Thus, historians of the time inform us that these views and actions were in response to "great earthquakes in divers places" and "stars in the sky [that] were seen throughout the whole world to fall towards the Earth, crowded together and dense, like hail or snowflakes. A short while later, a fiery way appeared in the heavens; and after another short period half the sky turned the colour of blood", consistent apparently with a swathe of suitably sized dust particles rapidly dispersing into the zodiacal cloud through their meteor stream. Similar but unexplained references to a former 'milky way' in the zodiac and a wine-coloured sea are also to be found in descriptions of the universe in much earlier accounts by the ancient Greeks (see Bailey *et al.* 1989). Thus although the difficult task of identifying where past Tunguskas may have exploded has not yet been undertaken – indeed we have been trained not to expect them! – the historical record is remarkably specific and the public response is not in doubt. A present day historian (*cf.* Riley-Smith, *loc. cit.*), for example, states that Urban's crusade and the nationwide church – building programmes towards the close of the eleventh century are best regarded as the climacteric of a temporarily restored general viewpoint according to which divine agencies would "taketh away kingdoms and changeth times". The belief that celestial agencies used 'weapons' to terminate nations, thereby transfering 'divine kingship' to the leader of a surviving nation, is a little appreciated but very old idea, reaching back to Sumero-Babylonian times.

Despite this striking evidence for a past association between sustained eschatological expectations and a probable Tunguska swarm, direct evidence of climatic recessions arising from encounters with such swarms is not yet established. Nevertheless the intersection at the time of Christ of the orbits of the Earth and Comet Encke, a member of the Taurid complex, (long after the violent break-up of the Encke progenitor around 3000 BC) has suggested to Whipple that this may be a significant period for meteor showers, and indeed he has provided evidence of a fragmenting body in much the same orbit only a few centuries after this time which could also be the eventual source of the eleventh century swarm. This suggests the fifth century may be especially important and the observation by Chinese astronomers of a "strange comet" in the same year (441) that Britain apparently experienced widespread destruction, followed by years of migration, darkened skies and a dark age (e.g. Myres 1986, Clube & Napier 1990), could have far greater significance in the context of multiple Tunguska encounters than modern historians have generally supposed. Thus the very first written reports of these times carry explicit references to "fire [that] fell from heaven" and that " did not die down until it had burned the whole surface of the island", the level of destruction and depopulation being such that many cities were still in a state of ruin a century later. This remarkable knowledge tends to be greeted with incredulity but we evidently cannot exclude the possibility that cultural dark ages are astronomically induced (in a physical as well as an eschatological sense). Indeed it may be noted that the author of these statements, who was chiefly concerned to admonish his contemporaries for general behavioural lapses, apparently did so some time shortly after alarming meteoric events in 524 that relate to an earlier fragmentation of the comet of 1845 which was also very close to the Earth in 441/2. The history of this comet, Comet Biela as it happens, and the Taurid stream appear to be of considerable interest therefore and it could well be that a seriously mistaken conception of these comparatively recent events contributes to the currently uncertain status of the theory of catastrophism. Obviously, a good deal more research in this area needs to be done: thus, in addition to the evidence for a climatic recession at this time, the evidence for wildfires and cosmic dust needs also to be sought.

That the demise of the dinosaurs and other extinctions have in the last decade turned out to have a significant bearing on our understanding of the past astronomical environment has already proved surprising enough. But that a single diatribe from a monastery in sub-Roman Britain at a particularly obscure period of European history should now have a bearing on the present astronomical environment may tax credulity to the limit. Nevertheless it should be noted that the obscurity in this instance does not exist for lack of evidence, it exists on account of the very serious difficulty that historians encounter in constructing a suitable motivating model of the period.

In this regard, by far the most curious fact is the extreme intensity of the historical reverse (within a period of admittedly growing decline) for it is generally agreed that the level of civilization previously enjoyed was not restored again for around 1300 years and of the time in question, there is no surviving written record for a century. A discontinuity dissected ceases to be a discontinuity but the events in question can justifiably be regarded as a catastrophe – certainly they are described as such in the first written record – and the course of events, involving reverses that are both sudden *and* steady, even has interesting parallels with those of the KT extinction, though on a greatly reduced scale! Terrestrial catastrophism may in fact be virtually uniformitarian on all timescales $\geq 10^3$ years!

To sum up, then, several areas of research have been identified on which theories of terrestrial catastrophism depend: (1) the true nature of spiral arms (i.e. do they contain massive amounts of dark matter? if so, how do they form?); (2) the episodic-cum-periodic nature of the geophysical record (i.e. is this its true form and how else could it be explained?); (3) the nature of the Earth-crossing asteroidal and meteoroidal complexes and the incidence of Tunguska swarms (i.e. do past and recent giant comets explain the asteroids and zodiacal cloud?); and (4) "fire from heaven" and the nature of the historical record (i.e. in their enthusiasm to secularise history, have modern historians simply removed the relevant astronomy along with the ancient divine influences?). It may be, of course, that the new environmental model has obvious very strong associations with biblical catastrophism, and that the latter has not been held in high regard during the present century. However, as we have seen, this state of affairs may simply reflect a failure to appreciate the role of giant comets.

Conclusion

The principal conclusion to emerge from the various lines of investigation discussed in this paper is that geological and biological evolution may well be dominated by the behaviour of successive giant comets in Earth-crossing orbits. The broad pattern of effects arising from each giant comet is an extended period of insolation first of all ($\sim 10^4$yr), resulting in a dense meteor stream and an 'ice-epoch' on Earth of corresponding intensity. In the aftermath of the ice-epoch, there is a more extended period of fragmentation ($\sim 10^4$-10^5yr) during which the dominant process on Earth is the input of submicron dust: this results in modulations of the atmospheric greenhouse, the principal frequency being that of precession of the orbital nodes. Paradoxically, however, the times of maximum input of submicron dust, when the Earth runs close to the core of the disintegrating giant comet, are just those times when the Earth is most at risk from a 'Tunguska-bombardment' resulting in armageddon-like conditions such as have been conceived for a so-called 'nuclear winter'. The climatic pattern includes a strong stochastic component therefore, and on geological timescales, may indeed reflect the input of even larger bodies of asteroidal (kilometre) dimensions. Even more paradoxically, though, it seems that giant comets can become virtually invisible during the fragmentation period, due to their immersion in obscuring dust, giving them overall something of the character of a 'holy grail'!

Acknowledgements

The material of this paper is based on a recent review to be published in *Contemporary Physics*.

Appendix

Our understanding of the universe and of galaxies owes nothing at present to astronomical inputs reaching the Earth and is based entirely upon an arbitrary view taken in 1930 as to the true nature of spiral arms. Thus it was imagined at this time that the material contents of all spiral arms in galaxies must be in a state of *continuous* circular motion, whilst also admitting the Oort-Lindblad theory of galactic rotation (Eddington 1930) and the Friedman-Lemaitre theory of universal expansion (Eddington 1931). Assuming continuous circular motion, the Oort-Lindblad theory provided typical values for galactic masses whence it followed, on the basis of the *only* general relativistic model of the stationary universe predicting the cosmological redshift (de Sitter 1917), that there was insufficient mass in the observable universe for the redshift to be explained in purely gravitational terms. Only the Friedman-Lemaitre theory remained therefore and it became necessary subsequently to accept the big bang theory of the expanding universe *and* the density wave theory of spiral structure.

However the assumption that spiral arms *must* be in a state of continuous circular motion is arbitrary. It is equally acceptable to suppose spiral arms represent a recurring or intermittent outflow from the centre (*cf.* Milne 1948, Ambartsumyan 1958), associated with the temporarily enhanced 'curvature' or inflated gravity of galactic nuclei (*e.g.* Jeans 1928). An extra, very large gravitational contribution to cosmological redshifts from short-lived, very massive phases in the evolution of galactic nuclei is not therefore excluded. On the other hand, the idea that pockets of inflated mass, with associated inflow and outflow, might appear and disappear in a stationary universe has received little further attention from cosmologists, leaving Eddington's arbitrary assumption to rule the astrophysical roost.

Recently, the discovery of unexpectedly large configurations of matter in the universe (*e.g.* Rubin & Coyne 1988) and of an unexpectedly distended disc devoid of unseen matter in our Galaxy (Kuijken & Gilmore 1989), whilst reducing the credibility of any proposed dynamical role for massive haloes, has not excluded the continuing action of intermittent pockets of mass inflation in galactic nuclei. Energy conservation, of course, requires an associated 'deflation' of the velocity of light to accompany the inflation of mass, implying also that various properties of matter must now depend in a very precise way on changes of gravitational potential (Dicke 1961, Atkinson 1962). Moreover, theory in this instance leads to a stationary universe, while the process of intermittent inflation in each galactic nucleus results in a correspondingly short-lived collapse of the host galaxy's central region followed by a *hot* spiral outflow (Clube 1988; *cf.* Bailey *et al.* 1989). Such outflows are expected to condense into highly differentiated giant comets *i.e.* the short-lived spiral arms contain massive amounts of dark matter in the form of "parent bodies" which combine to form stars and planets. The conventional understanding of galactic dynamics is modified as well in that galaxies are expected, on the stationary universe picture, to undergo violent relaxation not just during their original formation (*cf.* Lynden-Bell 1967) but repeatedly.

The significance of these active galactic nuclei, not perceived in Eddington's time but now apparently observed *in extremis* as quasars of high gravitational redshift, is their possible relevance so far as terrestrial catastrophism is concerned. Thus typical Oort cometary clouds are necessarily dispersed into the interstellar environment during the inflationary events in galactic nuclei and must subsequently be acquired from star-forming regions in spiral arms. Moreover, the captured giant comets that are then deflected into short-period orbits, by the very nature of their *hot* formation, have highly devolatilised, very fragile cores which, unlike comets with a conventional history of cold accretion, must undergo rapid fragmentation. This leads very naturally therefore to a general evolutionary state for the Earth which is neo-catastrophist *i.e.* the astronomical inputs to the Earth are discrete on a timescale of $\sim 10^5$yr (due to giant comets) *and* episodic-cum-periodic (due to the role played by massive spiral arms). These characteristics of the terrestrial record, which are surprisingly evident in the geomagnetic reversal frequency, have not yet been given a satisfactory alternative explanation and are therefore fortuitous within the framework of conventional cosmology and geophysics.

The idea that the wrong cosmology was introduced by Eddington in 1930 because evolution was not perceived at this time as being other than uniformitarian, only arose when it was realized that the disc of our Galaxy may be permeated by a strong spiral outflow (Clube 1973, 1978). The evidence for spiral outflow had not till then been regarded as particularly persuasive whilst further convincing evidence for its reality based on observations of the cirumnuclear disc at the Galactic centre has recently added further weight to the conclusion (Clube & Waddington 1989 and references therein). One should obviously be cautious in setting aside conventional theory because of its failings in respect of spiral outflow and terrestrial catastrophism, but it should also be kept in mind perhaps that the time interval between significant astronomical inputs on the basis of 'giant comet' theory (ie the Tunguska event of 1908 and the meteoroidal swarm of 1975) is now comparable to the survival time of received cosmology. The issue at stake is not entirely academic therefore!

References

Alvarez, L.W., Alvarez, W., Asaro, F. & Mitchel, H.V., 1980. *Extraterrestrial cause for the Cretaceous-Tertiary extinction*, Science, **208**, 1095-1108.

Alvarez, W. & Muller, R.A., 1984. *Evidence from crater ages for periodic impacts on the Earth*, Nature, **308**, 718-720.

Ambartsumian, V.A., 1958. *On the evolution of galaxies*, LA STRUCTURE ET L'EVOLUTION DE L'UNIVERS, 11th Solvay Conference on Physics, University of Brussels, ed. Stoops, R., 241-274.

Astapovic, I.S. & Terenteva, A.K., 1968. *Fireball radiants of the 1st-15th centuries*, Physics and Dynamics of Meteors, eds. Kresak, L. & Millman, P.M., IAU Symp. No. 33, 308-319. Reidel, Dordrecht, The Netherlands.

Atkinson, R. d'E., 1963. *General relativity in Euclidean terms*, Proc. R. Soc. London, Ser. A, **272**, 60-78.

Bailey, M.E. & Clube, S.V.M., 1978. *Recurrent activity in galactic nuclei*, Nature, **275**, 278-282.

Bailey, M.E., Wilkinson, D.A. & Wolfendale, A.W., 1987. *Can episodic comet showers explain the 30-Myr cyclicity in the terrestrial record?* Mon.Not.R.Astron.Soc., **227**, 863-885.

Bailey, M.E., Clube, S.V.M. & Napier, W.M. 1989 THE ORIGIN OF COMETS, (Pergamon), in press.

Baldwin, R.B., 1985. *Relative and absolute ages of individual craters and the rate of infalls in the Moon in the post-Imbrium period*, Icarus, **61**, 63-91.

Brecher, K., 1984. *The Canterbury swarm*, Bull.Amer.Astron.Soc., **16**, 476.

Brownlee, D.E., 1985. *Cosmic dust: collection and research*, Annu. Rev. Earth Planet. Sci., **13**, 147-173.

Butterfield, H., 1981. THE ORIGINS OF HISTORY. Methuen & Co., London.

Clube, S.V.M., 1973. *Another look at the absolute proper motions obtained from the Lick pilot programme*, Mon. Not. R. Astron. Soc., **161**, 445-463.

Clube, S.V.M., 1978. *Does our Galaxy have a violent history?* Vistas Astron., **22**, 77-118.

Clube, S.V.M., 1987b. *The origin of dust in the solar system*, Philos, Trans. R. Soc. London, Ser. A, **323**, 421-436.

Clube, S.V.M., 1988b. *Dust and star formation in a hot differentiating medium*, in DUST IN THE UNIVERSE, eds. Bailey, M.E. & Williams, D.A., 331-339. Cambridge University Press.

Clube, S.V.M. & Asher, D.J., 1989. *The proposed giant comet fragmentation ca. 2700 BC and its relationship with the α, β, γ zodiacal bands.* Paper presented at Asteroids Comets Meteors III, Uppsala 1989.

Clube, S.V.M. & Napier, W.M., 1984b. *The microstructure of terrestrial catastrophism*, Mon.Not.R.Astron.Soc., **211**, 953-968.

Clube, S.V.M. & Napier, W.M., 1986a. *Mankind's future: an astronomical view.* In COMETS, ICE AGES AND CATASTROPHES, Interdisciplinary Science Reviews, **11**, (No. 3) 236-247. J.W. Arrowsmith.

Clube S.V.M. & Napier, W.M., 1989a. *An episodic-cum-periodic galacto-terrestrial relationship*, Mon.Not.R.Astron. Soc., submitted.

Clube, S.V.M. & Napier, W.M., 1989b. COSMIC WINTER. Basil Blackwell Ltd., Oxford.

Clube, S.V.M. & Waddington, W.G. 1989a. *Velocities and the line-of-sight distibution of molecular clouds close to the Galactic Centre*, Mon. Not. R. Astron. Soc., **237**, 7P-13P.

De Sitter, W., 1917. *On Einstein's theory of gravitation, and its astronomical consequences*, Mon. Not. R. Astron. Soc., **78**, 3-28.

Dicke, R.H., 1962. *Mach's principle and equivalence*, EVIDENCE FOR GRAVITATIONAL THEORIES, ed. Moller, C., 1-49. (Proc. Internat. Sch. Phys. 'Enrico Fermi', Course XX.) Academic Press, New York.

Dohnanyi, J.S. 1978 *'Particle Dynamics'.* In COSMIC DUST (ed. J.A.M. McDonnell), pp 527-605, Chichester: Wiley.

Dorman, J., Evans., Nakamura, Y. & Latham, G., 1978. *On the time -varying properties of the linar seismic meteoroidal population*, Proc. Lunar Planetary Sci. Conf., **9**, 3615-3626.

Duncan, M., Quinn, T. & Tremaine, S.D., 1987. *The origin of short-period comets*, Astrophys. J. Lett., **328**, L69-L73.

Eddington, A.S. 1931 *'Council note on the expansion of the universe'* Mon.Not.R.Astron.Soc., **91**, 412-416.

Eddington, A.S. 1930 *'The Rotation of the Galaxy'* Halley Lecture, No 21 (Univ. of Oxford).

Everhart, E., 1972. *The origin of short-period comets.* Astrophys. Lett., **10**, 131-135.

Fechtig, H., 1982. *Cometary dust in the solar system*, COMETS, ed. Wilkening, L., IAU Coll. No. 61, 370-382, University of Arizona Press, Tucson.

Grun, E., Zook, H.A., Fechtig, H. & Giese, R.H., 1986. *Collisional balance of the meteoritic complex*, Icaurus, **62**, 244-272.

Hallam, A., 1984b. *Pre-quarternary sea-level changes.* Annu. Rev. Earth Planet. Sci., **12**, 205-243.

Hills, J.G., 1981. *Comet showers and the steady-state infall of comets from the Oort Cloud*, Astron. J., **86**, 1730-1740.

Hoyle, F. & Wickramasinghe, N.C., 1985. LIVING COMETS. University College Cardiff Press.

Hut, P., Alvarez, W., Elder, W.P., Hansen. T. Kauffman, E.G., Keller, G., Shoemaker, E.M. & Weissman, P.R., 1987. *Comet showers as a cause of mass extinctions*, Nature, **329**, 118-126.

Jeans, J.H., 1928. ASTRONOMY AND COSMOGONY. Cambridge University press.

Keynes, J.M. 1947. *Newton, The Man*, The Royal Society Newton Tercentenary Celebrations (15-19 July 1946), pp. 27-34. Cambridge University Press.

Kresak, L., 1978a. *The Tunguska object: a fragment of Comet Encke?*, Bull. Astron. Inst. Czechol., **29**, 129-134.

Kuijken, K. & Gilmore, G. 1989 *' The mass distribution in the Galactic disk. I,II,III'*. Mon.Not.R.Astron.Soc., in press.

Lindsay, J.F. & Srnka, L.J., 1975. *Galactic dust lanes and lunar soil*, Nature, **257**, 776-778.

Link, F., 1958. *Kometen, Sonnentatigkeit and Klimeschwankungen*, Die Sterne, **34**, 129-140.

Lynden-Bell, D. 1967 *'Statistical mechanics of violent relaxation in stellar systems'*. Mon.Not.R.Astron.Soc., **136**, 101-121.

Milne, E.A., 1948. KINEMATIC RELATIVITY, Clarendon Press, Oxford.

Myres, J.N.L., 1986. THE ENGLISH SETTLEMENTS, Clarendon Press, Oxford.

Napier, W.M. & Clube, S.V.M., 1979. *A theory of terrestrial catastrophism*, Nature, **282**, 455-459.

Oates, J., 1979. BABYLON. Thames & Hudson Ltd., London.

Rampino, M.R. & Stothers, R.B., 1984a. *Terrestrial mass extinctions, cometary impacts and the Sun's motion perpendicular to the galactic plane*, Nature, **308**, 709-712.

Raisbeck, G.M., Yion, F., Jouzel, J. & Petit, J.R., 1990. ^{10}*Be in polar ice cores as a probe of the solar variability influence on climate'*. Presented at Royal Society Discussion Meeting 15,16 February 1989, London; in press.

Rickman, H., 1986. *Masses and densities of comets Halley and Kopff*, In THE COMET NUCLEUS SAMPLE RETURN MISSION, ed. Melita, O., ESA SP-249, 195-205. ESA Publications, ESTEC, Noordwijk, The Netherlands.

Riley-Smith, J.S.C., 1986. THE FIRST CRUSADE AND THE IDEA OF CRUSADING. The Athlone Press, London.

Rubin, V.C. & Coyne, G.V., 1989 LARGE SCALE MOTIONS IN THE UNIVERSE. Princeton University Press.

Shoemaker, E.M., 1983 *Asteroid and comet bombardment of the earth*, Annu. Rev. Earth Planet. Sci., **11**, 461-494.

Singer, S.F. & Stanley, J.E. 1980 *'Submicron particles in meteor streams'*. In SOLID PARTICLES OF THE SOLAR SYSTEM (ed. I. Halliday & B.A. McIntosh), pp 329-332. Dordrecht: Reidel.

Smit, J. & Hertogen, J., 1980. *An extraterrestrial event at the Cretaceous-Tertiary boundary*, Nature, **285**, 198-200.

Sonett, C.P. 1990 *'The spectrum of radiocarbon variations'* Presented at the Royal Society Discussion Meeting 15,16 February 1989, London, in press.

Stohl, J., 1983. *On the distribution of sporadic meteor orbits.* In ASTEROIDS COMETS METEORS, eds. Lagerkvist. C.-I. & Rickman, H., 419-424. Uppsala Observatory, Uppsala, Sweden.

Sykes, M.V., Lebofsky, L.A., Hunten, D.M. & Low, F.J., 1986 *The discovery of dust trails in the orbits of periodic comets*, Science, **232**, 1115-1117.

Thomas, K., 1971. RELIGION AND THE DECLINE OF MAGIC. Weidenfeld & Nicolson, London.

Tian-shan, Zhuang, 1977. *Ancient Chinese records of meteor showers*, Chinese Astron. Astrophys., **1**, 197-220. (Translated from Acta Astron. Sinica, **14**, 37-58, 1966).

Toulmin, S., 1985. THE RETURN TO COSMOLOGY. University of California Press, Berkeley & Los Angeles.

Wetherill, G.W. & Revelle, D.O., 1982 *'Relationships between comets, large meteors and meteorites'*, In COMETS (ed. L.L. Wilkening), pp 297-319, Tucson: University of Arizona Press.

Wetherill, G.W., 1988. *Where do the Apollo objects come from?* Icarus, **76**, 1-18.

Whipple, F.L. & Hamid, S.E., 1952. *On the origin of the Taurid meteor streams*, Helwan Obs. Bull. No. 41, 1-30. Royal Observatory, Helwan. Fouad I University Press, Cairo. (Also see Whipple 1972d, 224-252).

Woolley, R.v.d.R., 1965. *Motions of the nearby stars*, GALACTIC STRUCTURE, eds. Blaauw, A. & Schmidt, M., 85-110. University of Chicago Press.

ON THE FREQUENCY OF REVERSALS OF THE GEOMAGNETIC DIPOLE

Kenneth M. Creer [1] and Poorna C. Pal [2]

[1] *Department of Geophysics, University of Edinburgh, Edinburgh EH9 3JZ, U.K.*

[2] *Instituto Astronomico e Geofisico, Universidade de Sao Paulo, 01051 Sao Paulo, SP Brazil*

Summary. The rate of occurrence of geomagnetic polarity reversals has varied widely through geological time. Accurate calibration of the polarity record via the sea floor magnetic anomaly pattern is possible only from the early Mesozoic. As documented for the past 165 Myr, the reversal record contains two mixed-polarity chrons separated by one of fixed (normal) polarity which occurred between about 80 and 110 Myr ago. During both mixed-polarity chrons, the reversal frequency oscillated about a secular trend, and spectral analyses have yielded periodicities of $\sim$ 15 Myr and $\sim$ 30 Myr, with the former being predominant through the current (post-Santonian) chron. The distribution of ages of mass extinction episodes, impact craters, tektite swarms *inter alia* can be arranged into 'bunches' which appear to be separated by intervals of $\sim$ 30 Myr, leading to suggestions that large bolides impacting on the Earth's surface might induce 'spurts' in reversal frequency. Theory provides for polarity reversals as a fundamental property of the geodynamo. The reversal rate, in the longer term, is probably controlled by intermittent (on the timescale of 10^8 years) convexion in the lower mantle. This provides a possible explanation for

the occurrence of the mixed- and fixed-polarity chrons. Over yet longer time-scales chrons can be grouped into super-chrons through which the polarity of the reversing dipole field is biassed in one sense rather than the other. The general view among geomagnetists is that if impacts have any effect at all on the operation of the geodynamo, it should be restricted to modulating reversal rates already prescribed by conditions existing at the core-mantle boundary at the time of the impact.

Introduction

The geomagnetic field is believed to originate from magnetohydrodynamic (MHD) processes in the Earth's fluid and electrically conducting outer core. Polarity reversals have occurred at irregular intervals throughout the geological past and they are endemic to the long term behaviour of the geomagnetic dynamo: see the discussions by Parkinson (1983), Jacobs (1984), Merrill & McElhinny (1983) and Krause & Roberts (1980). The observed lengths of individual normal (N) and reversed (R) polarity sub-chrons fit a Poisson statistical distribution (Cox 1969) so that there has been some reluctance, especially among geomagnetists, to accept cyclicity in the polarity reversal record.

The polarity reversal record

Calibration of time-scales. Since the pioneer study of Heirtzler *et al.* (1968), progressive steps have been taken to date the geomagnetic polarity reversal record. Several time-scales have been developed. Absolute ages are assigned by biostratigraphic zonation of stage boundaries. In the time-scale developed by Labreque *et al.* (1977, hereafter LKC), constant rates of sea floor spreading were assumed for the the South Atlantic during the Tertiary (back to $\sim$ 65 Myr ago) and for the North Pacific during the Maastrichtian and Campanian (from $\sim$ 65 to $\sim$ 75 Myr ago). This interval contains 191 reversals including sub-chrons as short as 40000 years.

Additional control points derived from magnetostratigraphic studies of Italian limestones were used to establish the Lowrie & Alvarez (1981, hereafter LA) time-scale which includes 174 reversals, fewer than the LKC time-scale since some of the short sub-chrons were omitted. There are 11 calibration points. Ages for the late Palaeocene — early Eocene are about 3 Myr younger than in the LKC time-scale and 1.8 Myr younger than in the time-scale of Ness *et al.* (1980). The reversal record can be extended back to 165 Myr by adding data for the Cretaceous fixed N polarity chron and then for the Oxfordian-Barremian mixed-polarity chron (from ~ 110 to ~ 165 Myr ago) as reconstructed by Lowrie & Ogg (1986).

An approximate 15 Myr periodicity. Mazaud *et al.* (1983) analysed the frequency of the reversal record covering the last 100 Myr. For both LKC and LA time-scales they identified an oscillating component which was superimposed on a monotonic trend. The latter could be represented either by a Lorenzian which peaks at the present time, or equally well by a straight line. The oscillating component, obtained by substracting the Lorenzian trend, was analysed by time autocorrelation. A frequency corresponding to a period of ~ 15 Myr was obtained for both the LA (14.1 — 15.8 Myr) and the LKC (13.9 — 15.4 Myr) time-scales for the post-Santonian mixed-polarity chron which extends back to ~ 80 Myr ago.

A purely poisson process? McFadden & Merrill (1984) found no indication of any periodicity in the reversal rate and hence came to the quite different conclusion that the reversal process is essentially a Poisson one. McFadden (1984a) argued that the identification of a ~ 15 Myr harmonic was due to the use of the fixed-length sliding window technique which can yield spurious frequencies when used to analyse synthetic random data. In reply to this criticism, Mazaud *et al.* (1984) carried out a further series of spectrum analyses using a selection of windows ranging in width from 1 to 10 Myr. They obtained periodicities confined to the narrow range 13 to 17 Myr and hence maintained that their original result was not due to a fortuitous choice of window length.

An approximate 30 Myr periodicity. Raup (1985) obtained a harmonic of ~ 30 Myr from spectrum analyses of 296 polarity reversals calibrated on the Harland *et al.* (1982) time-scale back to 165 Myr. It was in phase with the ~ 15 Myr periodicity already discussed. Raup used the minimum dispersion method which identifies frequencies present in a time-series by examining its correlation with a synthetic time-series containing known frequencies. Lutz (1985) questioned Raup's result on the basis of an analysis of the same data set, using a circular model rather than a linear one, and obtained 'apparent' periodicities which changed as the data set was truncated (the wrapping effect). Therefore he argued that Raup's 30 Myr periodicity was an artifact of the data length.

Later, Lutz & Watson (1988) presented an additional argument against accepting the ~ 30 Myr periodicity, maintaining that it could have been introduced by the inclusion of the Cretaceous fixed-polarity chron, even if it were not actually present in the data. Other criticisms were advanced by McFadden (1987), by Stigler (1987) and by Stigler & Wagner (1987).

Reversal "spurts". Figure 1 shows the reversal record back to 150 Myr ago, after smoothing with a fixed rectangular window of 5 Myr moved in 2.5 Myr steps. "Spurts" in reversal frequency at 15 Myr intervals are clearly visible to the eye in the post-Santonian segment and they seem to alternate in height. But only two "spurts" separated by ~ 30 Myr are seen in the Oxfordian-Barremian sequence (163 —

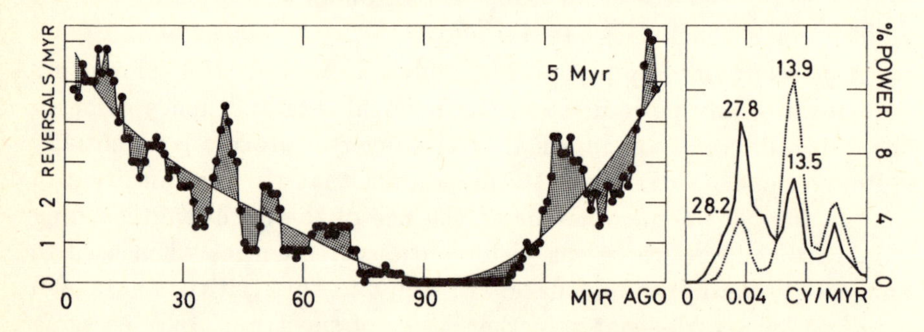

Figure 1. Geomagnetic polarity reversal frequency since Oxfordian times (0 — 165 Myr ago). The data were smoothed by a 5 Myr wide rectangular window and detrended. MEM (maximum entropy method) power spectra are shown on the right. The solid lines refer to the complete interval studied (0 — 165 Myr ago) while the dotted lines refer to time since the Cretaceous N chron (i.e. back to about 80 Myr ago). Periods corresponding to the MEM peaks are given in Myr.:

119 Myr ago). The corresponding MEM spectra, also shown in Figure 1, exhibit two peaks, at ~ 15 Myr and ~ 30 Myr, but the former is clearly the stronger through post-Santonian time while the latter is

the stronger for the Oxfordian-Barremian segment.

In order to test the validity of some of the already mentioned criticisms of the methodology of spectrum analysis applied to the reversal record, portions of the record of different lengths were analysed. The spectral peaks at $\sim$ 15 Myr and the $\sim$ 30 Myr remained robust through a wide range of segments of the record: for the post-Santonian chron, estimates of the $\sim$ 15 Myr period remained within 13.2 and 16.6 Myr for 0–30, 0–40, 0–50, 40–83, 20–83, 20–70, 20–60 and 0–83 Myr segments, and while the $\sim$ 30 Myr period was weak or absent in the first three of these segments, estimates varied between 26.5 and 30.6 Myr for the other segments.

On the use of fixed-chron-numbering windows. The secular trend observed in the frequency of polarity reversals through time leading into and following on from the Cretaceous N chron means that there must be a progressive change in the length of individual N and R chrons which appear in successive fixed-time windows. It follows that there will be a concomitant increase in variance. Therefore McFadden (1984b) proposed that windows containing a fixed number of polarity intervals should be used in order to obtain a constant variance.

To check the validity of this criticism, we reanalysed the data of Figure 1 using windows containing a fixed number (12) of polarity chron-pairs (N+R), moved in successive steps of two chron-pairs. We then computed the reversal frequency as the reciprocal of the mean interval length, following McFadden & Merrill (1984). As it turns out, the results, shown in Figure 2, are not very different from those for fixed-time windows, illustrated in Figure 1, showing that McFadden's criticism does not, in fact, invalidate results already obtained using fixed-time windows.

Again, it is clearly seen that most of the power is concentrated in the background trend of progressively decreasing reversal rate as the fixed polarity chron is approached from either end (Figure 3). When this trend is removed, the remaining power resides principally in two harmonics, at 27 – 33 and at 14 – 15 Myr. MEM and Fourier spectra for 0 — 165 Myr ago (the solid and dashed curves of Figure 4 respectively) show that the $\sim$ 30 Myr harmonic is stronger than the $\sim$ 15 Myr harmonic when the entire record is analysed. But the $\sim$ 15 Myr harmonic is seen to be stronger than the $\sim$ 30 Myr harmonic through the post-Santonian mixed-polarity chron (the dotted curve of

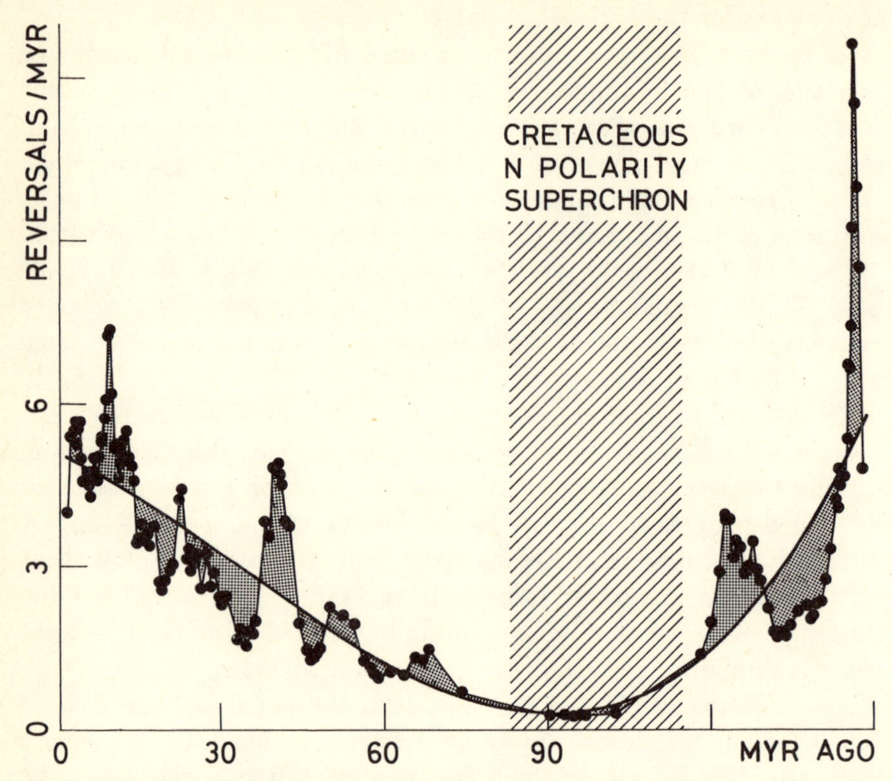

Figure 2. Reversal frequency variation for the data of Figure 1 when these reversals are seen through a fixed window of 12 (N+R) intervals, moved in steps of 2 (N+R) intervals. The reversal frequency is defined as the reciprocal of mean interval length. The secular trend and the detrended part of reversal frequency curves are shown:

Figure 4).

When variations in reversal frequency are examined as a function of interval number (Figure 5a), starting from the present, they appear to peak once every $\sim$ 40th interval (Figure 5b), as if they possessed a Markovian memory that lasted for about 40 intervals (Pal 1989). This result is equivalent to the postulation of periodicities.

Reversal frequency and extinctions

Alternative hypotheses. Mechanisms put forward for triggering extinc-

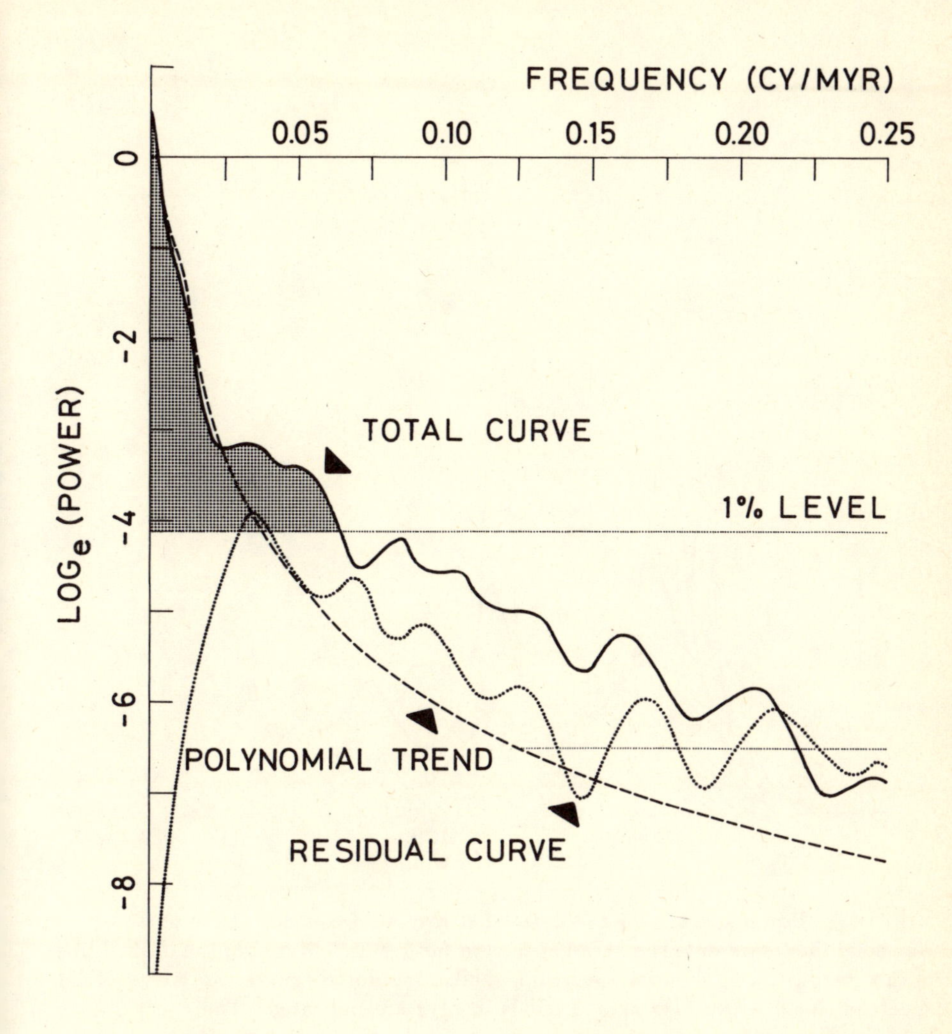

Figure 3. Fourier power spectra for the reversal frequency curves of Figure 2. The solid line is for the total curve, the dashed line is for the secular trend and the dotted line is for the detrended curve. Most of the power is contained in long wavelengths corresponding to the secular trend.:

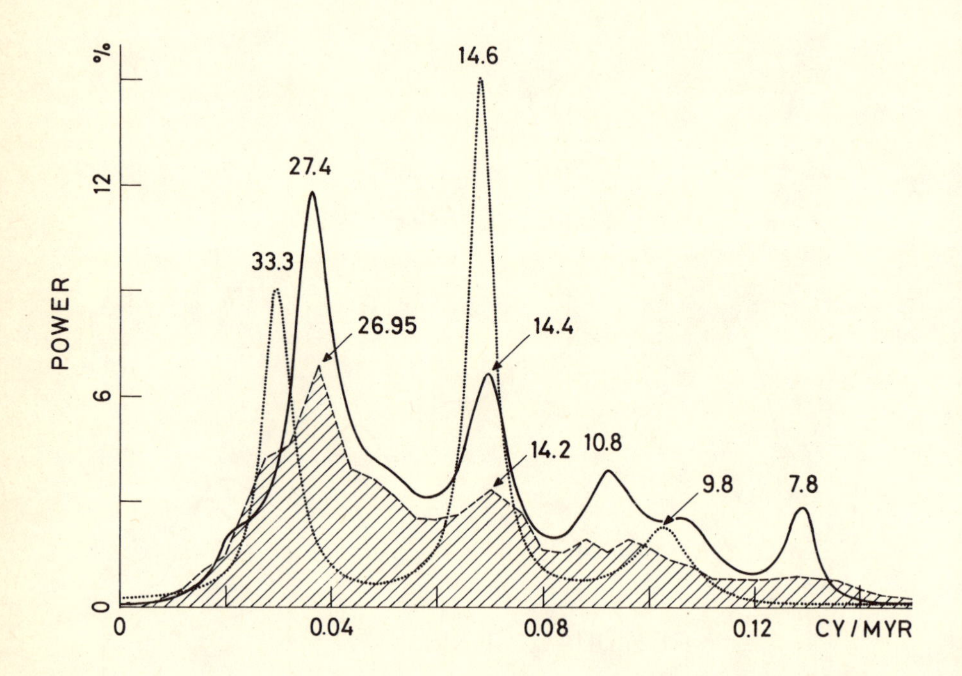

Figure 4. Power spectra for the detrended reversal frequency curve of Figure 2. The solid line represents the MEM spectrum for 0 — 165 Myr ago; the dashed line is the corresponding Fourier spectrum; whilst the dotted curve shows the MEM spectrum for 0 — 80 Myr ago. Periods in Myr are indicated. The spectra have been computed after linear interpolation at equal spacing of 1 Myr.:

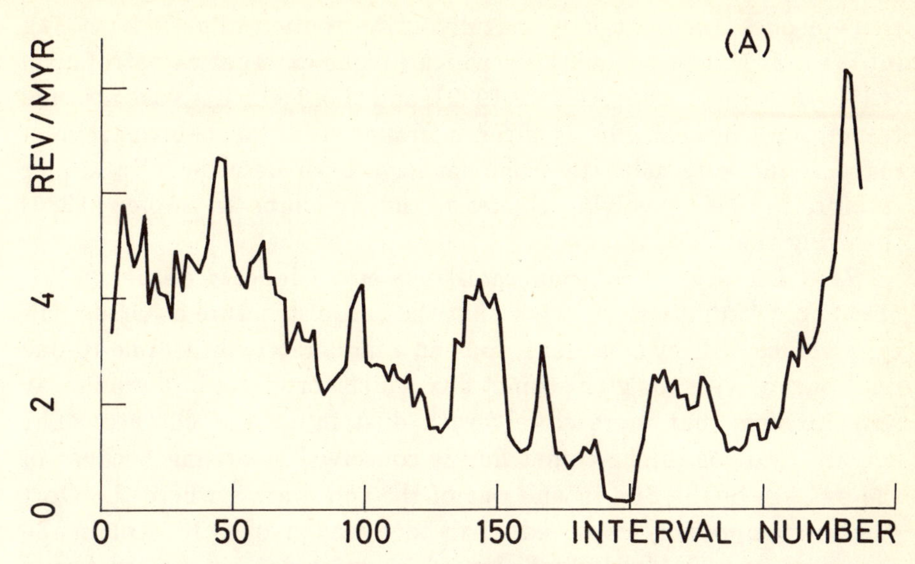

Figure 5(a). Reversal frequency, estimated as the reciprocal of mean interval length for 16 successive (N+R) intervals shifted by 2 (N+R) intervals each time, as a function of interval number (counted from the present).:

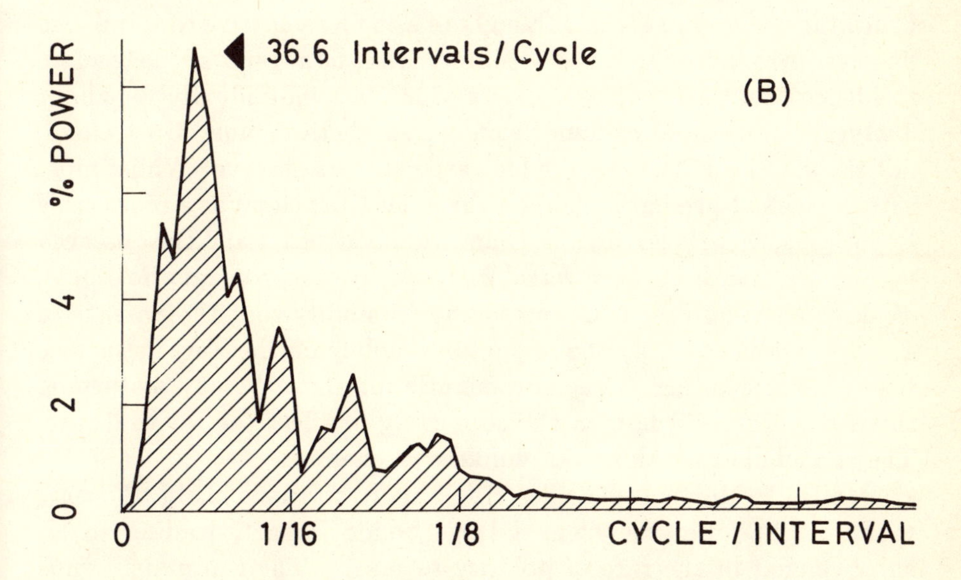

Figure 5(b). The Fourier power spectrum corresponding to Figure 5(a), suggesting that the reversal frequency peaks once every ~40 intervals:

tion episodes, though not specifically in the context of periodicity, fall into two distinct classes: those which invoke external catastrophism (Clube & Napier 1982; Alvarez 1983), and those which invoke internal cataclysms (McLean 1985; Officer & Drake 1985). Mechanisms involving periodic external catastrophism have been described by Muller & Morris (1986) involving Nemesis, and by Clube & Napier (1986) involving the Galaxy.

Models based on external catastrophism. Clube & Napier (1982, 1986) hold strongly to the view that the chron structure is astronomically induced. They consider it to be in a moderately direct one-to-one relationship with the giant comet flux which correlates in a crude way with impacts, but more closely with dust input and climate. The overall variation in the comet flux is conceived as arising because of the passage of the Sun in and out of the spiral arms where the Oort cloud is temporarily captured. This idea was invoked to explain the approximate 100 Myr episodicity which is evident in the cratering flux and the meteoroidal deposition records on the Moon as well as the terrestrial record. The key thing about this picture is that it also predicts the superchron structures and a prominent $\sim$ 15 Myr cycle.

In the theory advanced by Bailey *et al.* (1987), the Oort cloud and hence the comet flux are not thought of as being sensitive to spiral arm passage. Any action in this case simply arises as a result of the random incidence of intense comet showers which last individually for about 3 Myr. These showers come from a hypothetical inner Oort cloud, not the standard Oort cloud which astronomers observe. Whilst most astronomers at present look upon the inner Oort cloud as a reasonable hypothesis, there is no obvious connection between it and the observed cyclicity or trends in the reversal frequency of the geomagnetic dipole.

Rice & Creer (1989) discussed the possibility that the passage of a shock wave caused by the impact of a bolide on the Earth's surface might cause spalling at the core-mantle interface thereby disturbing the pattern of fluid flow in the core so as to affect the reversal rate. They conclude that this is an unlikely mechanism.

Muller & Morris (1986) describe a more tortuous chain of events which might follow on from a large bolide impact, leading to an enhancement of the rate of polarity reversals. Their argument runs as follows:– a large impact causes severe atmospheric pollution; this leads to climatic cooling which has to be sufficient to thicken the polar ice-cover; it follows that water will be transferred from mid and low latitudes to polar latitudes, so decreasing the Earth's moment of inertia; the consequent increase in angular velocity would be taken up initially by the mantle; hence the outer layers of the liquid core would be subjected to shear forces at the core-mantle boundary; these shear forces would affect the zonal pattern of fluid flow in the core associated with dynamo action. Muller & Morris say that this would enhance the frequency of reversals.

Acceleration of the angular velocity of the core would be *via* Ekman suction (Gubbins & Roberts 1987) in which process the the meridional (poloidal) flow is coupled to the geostrophic flow. Thus an increase in meridional flow should follow indirectly from the application of shear forces to the surface layers of the outer core. However, when the meridional circulation of the geophysically relevant alpha–omega dynamo is sufficiently large, the stationary mode is more readily excited by dynamo action than the oscillatory mode (Braginskii 1976; Roberts 1987). Therefore a reduction in reversal frequency is as likely to result from the Muller-Morris mechanism as the increase they proposed. McFadden & Merrill (1984) also suggest that a reduction in reversal frequency would result.

Models based on internal cataclysm. Loper & McCartney (1986) and Loper *et al.* (1988) suggested that an enhancement in reversal frequency might originate from intermittent break-up of the basal (D″) thermal boundary layer of the lower mantle. In this scenario,mass extinctions are attributed to environmental changes caused indirectly from cataclysmic mantle degassing, evidenced for instance, in flood basalt provinces such as the Deccan lavas of peninsular India. It is argued that whenever the thickness of the D″ layer builds up to a critical value, plumes are ejected from it, thereby reducing its thickness and hence its effectiveness in thermally insulating the core so as to enhance the reversal frequency. This would speed up convexion in the liquid core. However McFadden & Merrill (1986) have argued that such conditions occurred at the mantle-core boundary during fixed-polarity chrons i.e. they inhibited reversal activity. Nevertheless the intensity of the dipole field appears to have been stronger during chrons of fixed-polarity (Pal & Roberts 1988), in support of Loper & McCartney's model.

Do reversal spurts correlate with impact evidence? Pal & Creer

(1986) discussed the apparent cyclicity in reversal frequency in the light of the ~ 30 Myr harmonic reported to occur in the distribution of ages of mass extinction episodes (Raup & Sepkoski 1984, 1986) and of impact structures (Alvarez & Muller 1984; Grieve *et al.* 1986; Shoemaker & Wolfe 1986). Therefore they filtered out the 15 Myr harmonic in order to examine the ~ 30 Myr harmonic more closely. They then found that the "spurts" associated with the 30 Myr harmonic correlate approximately with the bunched age groups of extinction and impact age data. Also they drew attention to the fact that no "spurts" occurred during the Cretaceous fixed-polarity chron, although some would have been expected on a strict ~ 30 Myr time-table.

Polarity reversals and the cosmos

Longer term cyclicity. Negi & Tiwari (1983) carried out a Walsh spectrum analysis of the polarities (N, R or mixed) reported in the published literature on palaeomagnetic studies of world-wide Phanerozoic rock formations (*i.e.* relating to the last 570 Myr). They inferred a long term cyclicity, with reversal periods of 285, 114, 64, 47 and 34 Myr. The 285 Myr period was associated with the cosmic year (the period of complete revolution of the Solar System around the Milky Way galactic centre), and the shorter periods with regularities arising from the galactocentric radial motion, interaction of the spiral density wave with the galactic orbit, and solar oscillation in and out of the orbital plane.

In our view this result should be treated with some reservation because age control of the reversal stratigraphy for Palaeozoic time is poor and the record is incomplete. Continuous age control from sea floor magnetic anomaly profiles is not possible since virtually no sea floor older than Jurassic remains in existence, having been subducted into the mantle. The age data for the diverse rock formations studied by the different palaeomagnetic laboratories are thus judged to be inadequate for reliable spectrum analysis. This is especially true if the existence of periodicities as low as about 50 Myr is to be established with confidence, though it is perhaps just possible that a random distribution of errors in individual formation ages would allow the identification of the longest of the periodicities proposed, i.e. the one of 285 Myr.

Furthermore, there remains the quite separate difficulty that the 'magnetic' ages often differ from the true ages of both igneous and sedimentary rock formations. This problem has been inadequately addressed in many palaeomagnetic studies, especially those of a decade or more ago.

Polarity bias. On the longer time-scale (superchrons), for intervals of the order of several mixed- and fixed-polarity chrons, the polarity of the geomagnetic dipole seems to have been biassed, with one polarity more stable than the other. The first compilation of the integrated percentage of reversed and normal magnetization as reported in the published literature was made by Simpson (1966). Crain *et al.* (1969) computed a Fourier power spectrum based on these data, averaged over 15 Myr intervals, and observed periodicities of ~300 Myr and ~80 Myr. Subsequently Crain & Crain (1970) reanalysed the same data, obtaining periods of ~150 and ~40 Myr. Essentially a strong bias to the reversed polarity state was observed through the late Palaeozoic and to the normal polarity state through the Mesozoic.

McElhinny (1971) made an updated compilation of the published palaeomagnetic polarity data. Ulrych (1972) smoothed these with a ninth order polynomial and then averaged them over 25 Myr intervals before carrying out MEM spectrum analysis. He obtained periodicities of ~700 and ~250 Myr with errors estimated at about 20%. Later, Irving & Pullaiah (1976) obtained periodicities of 297, 113 Myr and 57 Myr (errors about 5 to 10%) from a MEM analysis of a new compilation of polarity data. Crain & Crain (1970) associated the polarity bias periodicity of ~ 300 Myr with the orbital period of the Solar System in the Galaxy. We have noted already, of course, that Negi & Tiwari (1983) subsequently inferred a similar association from an analysis of reversal frequency.

Creer (1975) examined the pattern of an independently compiled palaeomagnetic polarity data set in a different light. Rather than constructing a smooth curve through the plot of percentages of reversed polarity as a function of geological time through the Phanerozoic as the above mentioned authors had done, he argued that the data were such that it was just as reasonable to fit a set of straight lines to them. Three, possibly four, strong discontinuities are thus revealed, at ~425,

~375 Myr, ~ 225 Myr and ~ 55 Myr ago, the last two coinciding, within the estimated observational error, with the Permo-Triassic and the Cretaceous-Tertiary boundaries. Counts of growth rings in fossil shells (Scrutton & Hipkin 1973, Pannella 1972) allow estimates of the number of days in the year to be made. The resulting calculations of the rate of change of the length of the day, i.e. of the Earth's angular velocity, exhibit marked discontinuities at about 415 and 360 Myr, at ~ 245 Myr and ~ 65 Myr ago (Creer 1975), showing a remarkable coincidence with the observed discontinuities in geomagnetic polarity bias. It is tempting to speculate that the end of the Palaeozoic Era might have been a consequence of an impact event of similar severity to that postulated for the end of the Mesozoic Era by Alvarez (1983).

Reversal rate and geodynamo theory.

The dynamo mechanism. A basic difficulty is that the nature of the power sources which drive the fluid motions in the core have not been firmly identified, though several candidate sources have been considered in the published literature, for example: latent heat of growth of the inner core; radioactivity (potassium 40?) and chemical differentiation. This makes it difficult to be able to arrive at any firm conclusions as to the effectiveness of surface events (impacts?) or deep mantle events (D″ layer?) in perturbing the fluid motions in the core which are at the heart of the dynamo mechanism.

It is possible to draw some general conclusions about the geodynamo however, which may limit possibilities. We consider some of these as follows:

Dynamo oscillations. Mixed-polarity and fixed-polarity chrons have been explained theoretically in terms of the kinematic geodynamo with characteristic times of the order of 100 Myr (Roberts 1972, Krause & Radler 1980).

Effect of noise. Jacobs (1986) noted that experiments conducted by Crossley *et al.* (1986) on the self-reversing dynamo models of Rikitake (1958), Robbins (1977) and Olson (1983) suggest that fluctuations in reversal frequency might result from input of noise into the self reversing geodynamo mechanism. But the reversals so produced occur on time-scales intermediate between the electromagnetic (~10^4 yr)

and dynamic (~10^5 yr) time constants of the core (Roberts 1988), both of which are short compared to the average length of polarity chrons (~250000 yr) through the Tertiary; the reversal patterns do not match the observed statistical distribution of N and R chrons (Kono 1987) and they do not exhibit the equivalent of fixed-polarity chrons (Crossley *et al.* 1986). However there is no agreement as to how the noise might originate: Lund & Olson (1989) proposed that secular variation could be the source while Valet *et al.* (1986) claim that secular variations occur independently of polarity reversals.
Mantle convexion. Heat generated in the core is lost at the core-mantle boundary (CMB) and hence heat loss from the core should be modulated by changes in the rate of convexion in the mantle. Mantle convexion is thought to be intermittent, on time-scales which happen to be of the same order as the lengths of mixed- and fixed-polarity chrons. Thus it is tempting to think of possible models relating the two phenomena. McFadden & Merrill (1984) discussed the influence of convexion in the lower mantle on boundary conditions of the dynamo mechanism, arguing that more reversals occur in unit time when convexion is vigorous, *i.e.* when the temperature gradient in the outer core is high. This would mean that the long quiet (fixed-polarity) intervals would be associated with periods of high heat flow through the CMB.

Conclusion

Before it can become generally accepted among Earth scientists that the frequency of occurrence of such diverse phenomena as geomagnetic polarity reversals, faunal extinctions, cratering, geochemical anomalies, *etc.* are somehow related, firm proof must be provided that the respective postulated 'periodicities' really do exist and that they are in phase. As of now this has not been achieved.

Neither the evidence currently available nor the current state of dynamo theory permit a definitive conclusion as to whether the rate of geomagnetic reversals can be affected by either extra-terrestrial catastrophism (e.g. bolide impacts, astronomically induced global climatic recessions) or internal cataclysm (e.g. originating from the lower mantle). The current state of geodynamo theory does not permit an answer to the question whether either single large inputs (Clube & Napier 1986) or showers of small impacts (Bailey *et al.* 1987) would be capable of disturbing the magnetohydrodynamic processes in the core sufficiently to influence the reversal frequency, let alone provide an answer to the question as to which might be the more effective.

Supposing that the 15 or 30 Myr harmonics really do exist, it is still pertinent to ask whether that has anything to do with the physics of the geodynamo. This is because apparent periodicities are occasionally exhibited by probabilistic processes for limited intervals of time (Grebogi *et al.* 1987, Ito 1980). Thus we might expect that chaotic fluid motions in the Earth's liquid core could have developed apparently regular patterns of behaviour occasionally (on the geological time-scale). Hence it is possible that the only reasonably well dated segment of reversal record we have available for study may not be typical of the overall pattern of reversal polarity history.

References

Alvarez, L.W., 1983. *Experimental evidence that an asteroid impact led to the extinction of many species 65 million years ago.* Proc. Natl. Acad. Sci. U.S.A. **80**, 627–640.

Alvarez, W. & Muller, R.A., 1984. *Evidence from crater ages for periodic impacts on the earth.* Nature **308**, 718–720.

Alvarez, L.W., Alvarez, W., Asaro, F. & Michell, H.V., 1980. *Extraterrestrial cause for the Cretaceous-Tertiary extinction.* Science **208**, 1095–1108.

Bailey, M.E., Wilkinson, D.A. & Wolfendale, A.W. 1987. *Can episodic comet showers explain the 30 Myr cyclicity in the terrestrial record?* Mon. Not. R. astr. Soc. **227**, 863–885.

Braginskii, S.I., 1976. *On the nearly axially-symmetrical model of the hydrodynamic dynamo of the earth.* Phys. Earth planet. Int., **11**, 191–199.

Cox, A., 1969. *Geomagnetic reversals.* Science **163**, 237–245.

Clube, S.V.M. & Napier, W.M., 1982. *The role of the episodic bombardment in geophysics.* Earth planet. Sci. Lett. **57**, 251–262.

Clube, S.V.M. & Napier, W.M., 1986. *Galactic cycles and the terrestrial record.* In THE GALAXY AND THE SOLAR SYSTEM, ed. R. Smoluchowski et al., pp 260–285. Univ. of Arizona Press, Tucson.

Crossley, D., Jensen, O. & Jacobs, J., 1986. *The stochastic excitation of reversals in simple dynamos.* Phys. Earth planet. Int., **42**, 143–153.

Crain, I.K., Crain, P.L. & Plant, M.G., 1969. *Long period Fourier spectrum of geomagnetic reversals.* Nature **223**, 283.

Crain, I.K. & Crain, P.L., 1970. *New stochastic model for geomagnetic reversals.* Nature **228**, 39–41.

Creer, K.M., 1975. GROWTH RHYTHMS AND EARTH'S HISTORY eds. Rosenberg, G.D. & Runcorn, S.K., J. Wiley and Sons Ltd., London, pp 293–318.

Grebogi, C., Ott, E. & Yorke, J.A., 1987. *Chaos, strange attractors, and fractal basin boundaries in nonlinear dynamics.* Science **238**, 632–638.

Grieve, R.A.F., Sharpton, V.L., Goodacre, A.K. & Garvin, J.B., 1986. *A perspective on the evidence for periodic cometary impacts on earth.* Earth planet. Sci. Lett. **76**, 1–9.

Gubbins, D. & Roberts, P.H., 1987. *Geomagnetism*, ed. J.A. Jacobs, **2**, pp 1–183, Academic Press, London.

Harland, W.B., Cox, A. Llewellyn, P.G., Pickton, C.A.G., Smith, A.G. & Walters, R., 1982. A GEOLOGIC TIME SCALE, Cambridge University Press.

Heirtzler, J.R., Dickson, G.O., Herron, E., Pitman, W.C. & Le Pichon, X., 1968. *Marine magnetic anomalies, geomagnetic field reversals and motions of the ocean floor and continents.* J. geophys. Res., **73**, 2119–2136.

Irving, E. & Pullaiah, G., 1976. *Reversals of the geomagnetic field, magnetostratigraphy and relative magnitude of palaeosecular variation in the Phanerozoic.* Earth Sci. Rev. **12**, 35.

Ito, K., 1980. Earth planet. Sci. Lett. **51**, 451–456.

Jacobs, J.A., 1984. REVERSALS OF THE EARTH'S MAGNETIC FIELD, Adam Hilger, Bristol.

Jacobs, J.A., 1986. *Reversals — from the core or the skies?* Nature **323**, 296–297.

Kono, M., 1987. *Rikitake two-disk dynamo and paleomagnetism.* Geophys. Res. Lett. **14**, 21–24.

Krause, F. & Radler, K.H., 1980. MEAN FIELD MAGNETOHYDRODYNAMICS AND DYNAMO THEORY, Pergamon Press, Oxford.

Krause, F. & Roberts, P.H., 1980. *Strange attractor character of large-scale non-linear dynamos.* Adv. Space Phys, **1**, 231–240.

Labreque, J.L., Kent, D.V. & Cande, S.C., 1977. *Revised magnetic polarity time scale for Late Cretaceous and Cenozoic time.* Geology **5**, 330–335.

Loper, D.E. & McCartney, K., 1986. *Mantle plumes and the periodicity of magnetic field reversals.* Geophys. Res. Lett., **13**, 1525–1528.

Loper, D.E., McCartney, K. & Buzyna, G., 1988. *A model for correlated epidocity in magnetic-field reversals, climate, and mass extinction.* J. geol., **96**, 1–15.

Lowrie, W. & Alvarez, 1981. *One hundred million years of geomagnetic polarity history.* Geology **9**, 392–397.

Lowrie, W. & Ogg, J.G., 1986. *A magnetic polarity time scale for the early Cretaceous and late Jurassic.* Earth Planet. Sci. Lett. **76**, 341–349.

Lund, S. & Olsson, P., 1987. *Historical and paleomagnetic secular variation and the earth's core dynamo process.* Rev. Geophys. **25**, 917–928.

Lutz, T.M., 1985. *The magnetic record is not periodic.* Nature **317**, 404–407.

Lutz, T.M. & Watson, G.S., 1988. *Effects of long-term variation on the frequency spectrum of the geomagnetic reversal record.* Nature **334**, 240–242.

Mazaud, A., Laj, C. de Seze, L. & Verosub, K.L., 1983. *15-Myr periodicity in the frequency of geomagnetic reversals since 100 Myr.* Nature **304**, 328–330.

Mazaud, A., Laj, C. de Seze, L. & Verosub, K.L. 1984, *Reply to P.L. McFadden.* Nature **344**, 396.

McElhinny, M.W., 1971. *Geomagnetic reversals during the Phanerozoic.* Science **172**, 157.

McFadden, P.L., 1984a. *~15 Myr periodicity of geomagnetic reversals since 100 Myr.* Nature **344**, 396.

McFadden, P.L., 1984b. *Statistical tools for the analysis of geomagnetic reversal sequences.* J. geophys. Res., **89**, 3363–3372.

McFadden, P.L., 1987. *Periodicity of the earth's magnetic reversals.* Nature **330**, 27.

McFadden, P.L. & Merrill, R.T., 1984. *Lower mantle convection and geomagnetism.* J. Geophys. Res., **89**, 3354–3362.

McFadden, P.L. & Merrill, R.T., 1986. *Geodynamo energy source constraints from palaeomagnetic data.* Phys. Earth planet. Int., **43**, 22–33.

McLean, D.M., 1985. *Deccan trap mantle degassing and the terminal Cretaceous marine extinctions.* Cret. Res., **61**, 235–259.

Merrill, R.T. & McElhinny, M.W., 1983. THE EARTH'S MAGNETIC FIELD, Academic Press, London pp 401.

Muller, R.A. & Morris, D.E., 1986. *Geomagnetic reversals from impacts on the earth.* Geophys. Res. Lett., **13**, 1177–1180.

Negi, J.G. & Tiwari, R.K., 1983. *Matching long term periodicities of geomagnetic reversal and galactic motions of the solar system.* Geophys. Res. Lett. **10**, 713–716.

Ness, G., Levi, S., & Couch, R., 1980. *Marine magnetic anomaly time-scales for the Cenozoic and Late Cretaceous: a précis, critique and synthesis.* Rev. Geophys. and Space Phys., **18**, 753–770.

Officer, C.B. & Drake, C.L., 1985. *Terminal Cretaceous environmental events.* Science, **227**, 1161–1167.

Olson, P., 1983. *Geomagnetic polarity reversals in the turbulent core.* Phys. Earth planet. Int. **33**, 260–274.

Pal, P.C., 1989. GEOMAGNETISM AND PALAEOMAGNETISM, ed: Lowes, F.J., Collision, D.W., Parry, J.H., Runcorn, S.K., Tozer, D.C. & Soward, A., Kluwer Academic Publishers, Dordrecht pp 319–334.

Pal, P.C. & Creer, K.M., 1986. *Geomagnetic reversal spurts and episodes of extraterrestrial catastrophism.* Nature **320**, 148–150.

Pal, P.C. & Roberts, P.H., 1988. *Long-term polarity stability and strengths of the geomagnetic dipole.* Nature **331**, 702–705.

Pannella, G., 1972. *Palaeontological evidence on the Earth's rotational history since the early Cambrian.* Astrophys. and Space Sci, **16**, 212–237.

Parkinson, W.D., 1983. INTRODUCTION TO GEOMAGNETISM, Scottish Academic Press, Edinburgh pp 433.

Raup, D.M., 1985. *Magnetic reversals and mass extinctions.* Nature **314**, 341–343.

Raup, D.M. & Sepkoski, J.J., 1984. *Periodicity of extinctions in the geologic past.* Proc. Natl. Acad. Sci. U.S.A. **81**, 801–805.

Raup, D.M. & Sepkoski, J.J., 1986. *Periodic extinction of families and genera.* Science **231**, 833-836

Rice, A. & Creer, K.M., 1989. GEOMAGNETISM AND PALAEOMAGNETISM, eds. Lowes, F.J., Collinson, D.W., Parry, J.H., Runcorn, S.K., Tozer, D.C. & Soward, A., Kluwer Academic Publishers, Dordrecht, pp 227–231.

Rikitake, T., 1958. *Oscillations of a system of disk dynamos.* Proc. Camb. Phil. Soc. **54**, 89–105.

Robbins, K.A., 1977. *A new approach to subcritical instability and turbulent transitions in a simple dynamo.* Math. Proc. Camb. Phil. Soc. **82**, 89–105.

Roberts, P.H., 1972. *Kinematic dynamo models.* Phil. Trans. R. Soc. Lond., **A272**, 663–703.

Roberts, P.H., 1987. *Geomagnetism* (Ed: J.A. Jacobs), **2**, 251–306. (Academic Press, London).

Roberts, P.H., 1988. *Geomagnetic polarity reversals: are they spontaneous or are they forced?* EOS Trans. Am. Geophys. U., **69**, 1437.

Scrutton, C.T. & Hipkin, R.G., 1973. *Long term changes in the rotation rate of the Earth.* Earth Sci. Rev. **9**, 259–274.

Shoemaker, E.M. & Wolfe, R.F., 1986. THE GALAXY AND THE SOLAR SYSTEM ed: R. Smoluchowski et al., Univ. of Arizona Press, Tucson pp 338–386.

Simpson, J.F., 1966. *Evolutionary pulsation and geomagnetic polarity.* Geol. Soc. Amer. Bull., **77**, 197.

Stigler, S.M., 1987. *Aperiodicity of magnetic reversals?* Nature **330**, 26.

Stigler, S.M. & Wagner, M.J., 1987. *A substantial bias in nonparametric tests for periodicity in geophysical data.* Science **238**, 940–945.

Ulrych, T.J., 1972. *Maximum entropy power spectrum of long period geomagnetic reversals*, Nature **235**, 218.

Valet, J.P., Laj, C. & Tucholka, P., 1986. *High resolution sedimentary record of a geomagnetic reversal.* Nature **322**, 27–32.

TERRESTRIAL CATASTROPHISM AND GALACTIC CYCLES

W.M. Napier

Royal Observatory, Blackford Hill, Edinburgh EH9 3HJ, U.K.

Summary. The origin of theories of terrestrial catastrophism is described with particular reference to the immediate environment of the Earth, periodicities in the terrestrial record and the likely galactic origin of these periodicities. The alternative proposal involving a hypothetical companion star to the Sun is also briefly reviewed.

Introduction

The first suggestion in modern times that a collision between the Earth and a celestial body might have had a catastrophic effect on terrestrial life appears to have been made by the palaeontologist de Laubenfels in 1956. Looking at the pattern of survival and extinction at the Cretaceous-Tertiary boundary, de Laubenfels was led to suggest that hot winds from a giant meteorite might have caused the famous dinosaur extinctions. Aquatic animals (apart from those which had to surface to breathe air) survived because water protected them from the intense heat; vegetation survived because of the regenerating power of roots and seeds; while birds and mammals survived because they were found at high, snow-covered latitudes: 'Even boiling hot air, blowing over miles of snow, would cool down to a breathable degree'. De Laubenfels thus added catastrophic impact to the fifty or so hypotheses which have been advanced to account for the great extinction. Unfortunately no estimate of the impact rate of 'giant meteorites' was made, and with no indication of whether the postulated impact was a plausible event, the suggestion was inevitably speculative.

Two years later, in a short article described as an 'Abstract in advance of publication' (although the publication never appeared), the Estonian astronomer Öpik computed the probability of impact of the Earth by active comets and Earth-crossing asteroids. He evaluated the area of devastation of such an impact by assuming the lethal effect to require 2000 seconds of heat action from volatilized rock streaming above the top of the local atmosphere. At the time Öpik wrote, over 30 years ago, the astronomical environment of the Earth was poorly understood, and his estimates of both impact rate and lethal range of impact were much too small: for example he calculated that impacts with bodies greater than 8.5 km in diameter would take place only once in 260 million years, and that the lethal area would be only 3.6 million kilometres, less than one percent of the area of the Earth. Öpik did not associate any particular mass extinction with an impact, but he did propose that the development of land life during the Proterozoic might have been inhibited by catastrophic collisions.

Such ideas, precursors of the modern studies of terrestrial catastrophism, are to be found scattered throughout the literature of the 1960's and 1970's: for example McLaren, in a presidential address to the Paleontological Society of America in 1970, suggested that a giant meteorite crashing into the Palaeozoic Pacific might have been responsible for generating intense turbidity in the ocean, choking filter-feeding creatures, and he associated one of the major marine mass extinctions with such an event; and Urey in 1973 proposed that a cometary collision would create a high-temper ıture pulse perhaps coupled with increased humidity, such an impact being capable of destroying *inter alia* the air-breathing marine dinosaurs. Urey considered it to be 'possible and even probable that a comet collision with the Earth destroyed the dinosaurs and initiated the Tertiary division of geologic time.'

A conspicuous feature of these and similar studies (*e.g.* Gallant 1964) is that either no attempt was made to assess impact probabilities or, in the work of Öpik, the derived impact rates were too small to be interesting. So long as there was nothing substantial to go on with regard to expected collision rates, the impact hypothesis was bound to be speculative and it had little influence on the mainstream of geological thought. The latter was in any case, from the mid 1960's, in the throes of the plate tectonic revolution. It may also be said of these early ideas that they did not strictly amount to theories of terrestrial catastrophism; rather, they described occasional, dramatic irruptions into an Earth basically controlled by uniformitarian processes.

That there might be a more continuous rôle for cosmic forces, although no viable mechanism was ever proposed, was implicit in another strand running through the geological literature, this time going back sixty years; this is the persistent claim that many geological phenomena occur in cycles. The astronomical connection lies in the fact that the claimed terrestrial cycles were recognised as having frequencies corresponding to natural galactic ones. A 30 million year cycle, which is about the interval between successive crossings of the Sun through the plane of the Galaxy, was discussed for example by Holmes (1927) in connection with sea level variations and orogenies, by Dorman (1968) in connection with global climatic variations; by Steiner & Grillmair (1973) in connection with glaciations; by Ager (1975), again in connection with sea level variations; by Fischer & Arthur (1975) in relation to marine extinctions; by Innanen *et al.* (1978) who noted a near coincidence of geological periods in general with the solar vertical oscillation; and by Negi & Tiwari (1983) in connection with the frequency of geomagnetic reversals. It is interesting that the 30 Myr cycle seems to have been claimed for each of these phenomena without reference to its discovery in the others, an independence which might tend to strengthen confidence that a real signal was being detected.

Longer cycles of 200-250 million years were also claimed for all of these phenomena. McIntyre (1977), for example, claimed that the extrusion of carbonatite followed a 230 Myr periodicity, Negi & Tiwari (1983) claimed a 285 Myr periodicity in the occurrence of mixed magnetic intervals, and so on: a review of the literature to 1980 on long-term fluctuations in the evolution of the Earth is given by McCrea (1981). While attention was frequently drawn to the fact that these longer periods were about that of the galactic year, the time taken for the Sun to orbit the centre of the Galaxy, once again no firm mechanism by which the Galaxy might affect the affairs of the Earth was ever proposed. After all, how could the Galaxy, which has a mean density about that of a hard vacuum, affect say the level of the oceans, the outpourings of volcanoes, or the geomagnetic dynamo, the latter separated from the galactic vacuum by 2,500 km of overlying mantle with twice the tensile strength of steel?

Occasionally it was suggested that supernova explosions or penetrations of very dense nebulae might be responsible for specific mass extinctions (Russell 1979) or ice ages (Shapley 1921, Hoyle & Lyttleton 1939, McCrea 1975), and other astronomical mechanisms which might cause prompt biological damage were also proposed from time to time (*e.g.* Wdowczyk & Wolfendale 1977). However these postulated mechanisms were too rare, too speculative or too circumscribed to be taken seriously as major determinants of Earth history; for example they could not account for the periodicities in the geological record, if these were indeed real.

To summarise, the position until about a decade ago was that, while suspicions were from time to time voiced that the astronomical environment might have an effect on terrestrial processes, either through occasional great catastrophes or some unspecified more continuous interaction, there was not enough hard evidence to make such a proposition stick. The lack of a working astronomical mechanism, the great uncertainty in bolide impact rates, the lack of rigour in the derivations of periodicity, and the paucity of the geological data, all combined to ensure that there was no serious challenge to the prevailing uniformitarian ethos.

Nevertheless, throughout the 1970's new discoveries were being made, mainly in astronomy, which were to radically change this situation. The application of new technologies was leading to the conclusion that both the environment of the Earth and that of the Solar System were much more 'active' places than had been realised. As a result, it began to appear that a coherent astronomical framework for a theory of terrestrial catastrophism could after all be constructed. The connecting link between Earth and Galaxy was found to be the Oort comet cloud, which is susceptible to galactic perturbations on the one hand, and a major supplier of bolides to the inner Solar System on the other. It was realised that episodes of disturbance of the Oort cloud would lead to episodes of bombardment of the Earth by cometary debris, recurring on timescales dictated by galactic processes (Napier & Clube 1979 *cf.* Clube 1978). In the decade since this broad proposition was first put forward, more information has become available about the nature of the terrestrial and Solar System environments, and the original hypothesis has been refined and updated.

The purpose of this article is first, to describe briefly the known astronomical environment and argue that it is one within which Earth disturbances, controlled by the Galaxy, are expected for a wide range of circumstances; and secondly, to argue that there does indeed seem to be a uniquely galactic periodicity in the terrestrial record. Basic terrestrial phenomena are thus controlled by a galactic 'clock'. There has also appeared, in the marketplace of ideas, the proposition that cycles of extinction might arise because the Sun has a companion star with a period coincidentally equal to the ~30 Myr interval between galactic plane crossings (Whitmire & Jackson 1984; Davies *et al.* 1984). This hypothesis does not arise as a natural consequence of the observed astronomical environment, but is rather an *ad hoc* device constructed to account for the 30 Myr cycle. Consideration will be given to its merits.

Finally in this introduction, some words on methodology may be appropriate. The approach adopted here is deductive: starting with the known astronomical environment, certain predictions are made and tested against terrestrial data. Thus, contrary to frequent statements made in the geoscience literature (*e.g.* Raup 1986), there is no question of inventing, *post hoc*, astronomical explanations to account for geochemical anomalies or periodicities in the terrestrial record. Indeed the basic galactic mechanisms, involving for example bombardment episodes as a cause of the KT extinctions and the prediction of galactic periodicities in the terrestrial record, was in the literature, at least in outline, before the KT iridium discoveries or the modern periodicity discussions (Clube 1978, Napier & Clube 1979; see also Clube & Napier 1982 a,b). Likewise, the common assumption (e.g. Hallam 1986) that one invokes causes external to the Earth for say the dinosaur extinctions as a sort of *deus ex machina* only after all else fails, is methodologically incorrect because it involves ignoring evidence, albeit evidence relating to the Earth's environment. The Earth is a planet, subject to external influences, and in determining its evolution a holistic view is of course required.

The environment of the Earth

That a temporary cessation of sunlight, due to stratospheric dust input, might have had a catastrophic effect on life at the end of the Palaeozoic was first suggested by Hoyle & Wickramasinghe (1978). These authors proposed that the Earth might have scooped up sufficient dust for the purpose during a passage through the tail of an exceptionally large comet. The encounter rate they derived was $\sim 10^{-8}\ \mathrm{yr}^{-1}$; however this seems to be an overestimate by one or two powers of ten for the exceptional comet required, and the encounter had to be so close that a collision was then in any case likely.

The proposition that the impact of a ~10 km, bolide raised dust into the stratosphere and blocked sunlight, thereby reducing photosynthesis and contributing to the extinction of the dinosaurs, was originally advanced by Napier & Clube (1979) as part of a new general theory linking major aspects of terrestrial evolution to galactic and spiral arm modulations of the comet flux from the Oort cloud (cf. Clube 1978, Clube & Napier 1982a,b, 1984a,b; 1986). Straightforward blast from a huge impact was considered to be possibly the prime effect, being strong enough for example to flatten forests worldwide; while the indirect effects of NO creation leading to ozone depletion and an ultraviolet deluge were also considered. Ocean impacts were found to yield continental shelf waves up to 8 km high.

The basis of the theory was a new quantitative estimate of the mean collision rate with Earth-crossing bodies of various masses (Table 1), coupled with the discovery of a galactic mechanism which would modulate that collision rate. These in turn derived from several, then recent, astronomical discoveries: (i) relatively precise data on the ambient small-body population and the terrestrial impact cratering history; (ii) the existence of a system of massive nebulae in the Galaxy, in whose presence the classical Oort comet cloud was shown to be unstable; and (iii) a possible ~50-100 Myr episodicity in the terrestrial and lunar records, the latter evidenced for example by variations in micrometeoroidal deposition on the Moon (Lindsay & Srnka 1975). This time interval was taken to be comparable to that between crossings of spiral arms.

Table 1. Mean interval δt between impacts with planetesimals of diameter $\geq$ d with impact energies $\geq$ E and producing craters of diameter $\geq$ D. The actual impact frequency is expected to be modulated by the galactic environment, with prolonged bombardment episodes occurring preferentially during crossings of spiral arms (~ 50–100 Myr), modulated by sharper bombardments caused by out-of-plane solar oscillations (~ 30 or 15 Myr cycles).:

δt (Myr)	d (km)	E (Mt)	D (km)
9	4	7	80
14	5	15	100
58	11	160	200
360	30	3,600	500

The argument was that the Oort comet cloud must be episodically disturbed on timescales ~ 50 Myr, and the flux of comets therefrom episodically enhanced, as the Sun passes through spiral arms. These enhancements lead to catastrophic episodes of bombardment of the Earth by the degassed remnants of comets. For the latter, following Opik (1963), it was assumed that an appreciable proportion of the population of Earth-crossing asteroids was cometary in origin. It was further demonstrated that the stresses acting on the Oort cloud were sufficiently intense that the Oort cloud could not have survived for the age of the Solar System, and it was assumed that replenishment would take place during passage through the molecular clouds in spiral arms, which were thus regarded as 'comet factories'. The flux enhancements were considered to be modest (factors of 3 or 4) although occasional surges up to ~ 10 times background were expected (Clube & Napier 1982a). The expected disturbances were sufficiently severe that, it was anticipated, geological as well as biological processes would be affected, yielding a tendency for 'magnetic field reversals, principal plate movements, ice ages and mass extinctions to be correlated and to recur on time scales dictated by galactic processes'.

By a curious turn of events, within a few months of publication of this hypothesis, the same proposition regarding the impact of a ~ 10 km bolide was presented almost simultaneously by several geological groups: by Alvarez *et al.* (1980) as an explanation of their discovery of excess iridium and osmium at the KT boundary, by Smit & Hertogen (1980) to account for an apparently abrupt planktonic extinction at the boundary, and by Hsu (1980) to account for the major KT marine extinctions. There was thus a partial convergence between conclusions drawn from the new astronomical data, and from new geological data at the KT boundary.

The scenario of the geologists was however once again in the mould of an isolated great disturbance imposed on a world dominated by uniformitarian processes, whereas the astronomical picture called for a closer control over terrestrial processes, induced by bombardment episodes and modulated by galactic patterns and periodicities. Amongst the testable predictions, a firm confirmation of galactic periodicities in the terrestrial record was an obvious candidate, evidence for multiple bombardment was another, and correlations between impact craters and geological phenomena or mass extinction horizons was a third.

Periodicities

In 1984, Raup & Sepkoski rediscovered the Fischer-Arthur cycle in the marine fossil record (although with period $\sim$26 Myr rather than 32 Myr). The KT extinctions fitted on to the cycle, and a spin-off from the result was the invention of the Nemesis hypothesis. A 'cottage industry' of cycle-hunting also followed, and 30 Myr cycles were then claimed to occur also in the geomagnetic reversal, tectonic and impact cratering records. All of these claims were rediscoveries of earlier neglected results, and all of them were disputed.

Even before the Raup-Sepkoski analysis, significant periodicities of not only 32-34 Myr (Negi & Tiwari 1983) but also of 15 Myr (Mazaud *et al.* 1983) had been claimed for the geomagnetic reversal record. The 15 Myr cycle, whose presence can also be seen in earlier data (*e.g.* Lowrie & Alvarez 1983), evidently complicates the situation: the geomagnetic reversal record is the most complete of all those studied, and Mazaud *et al.* considered that their cycle was 'one of the most persistent long-term periodicities ever reported for a geophysical phenomenon'. If there were a 15 Myr cycle in this record, imposed by periodic bombardments affecting the other phenomena, it should exist in these other phenomena too, perhaps having gone unnoticed.

In fact, there are indications of a 15 Myr cycle in many of the data which have been put forward to support a ~30 Myr one. Raup (1985) found a ~30 Myr geomagnetic reversal periodicity precisely in phase with the 15 Myr one and was unclear which period was a harmonic of the other, and similar 'harmonics' can be seen in the geomagnetic reversal study of Stothers (1986). Grieve *et al.* (1985), from a study of 26 craters $\geq$ 5 km across with well-determined ages, found that a number of periods could be derived in the range 15-20 Myr and 30-35 Myr depending on the precise data selection criteria. Shoemaker & Wolfe (1986), from a very similar data set, found a best-fit periodicity in the range 30-32 Myr, with the most recent peak occurring 0-3 Myr ago; however an examination of their paper reveals that over the last 100 Myr the data are better fitted by a ~15 Myr cycle. Pandey & Negi (1987) claimed that major global volcanic episodes over the last 250 Myr showed a ~33 Myr cycle, but their data clearly reveal three intermediate episodes for the last 100 Myr, yielding an overall ~16 Myr cycle.

It seems on the face of it, therefore, that there is a tendency to look for a 30 Myr period while ignoring evidence for a 15 Myr one: subjectivity or preconception may have been at work in some of these analyses (see also the cautionary remarks by Heisler & Tremaine (1989) regarding subjectivity in the choice of data). A coherent and rigorous approach to this controversial field seems to be called for. In this Section such a technique is discussed in outline, and then applied to various terrestrial data. A more complete discussion is given elsewhere (Clube & Napier 1989).

Technique. Consider a set of N signals, each associated with a time. Following Mardia (1972) and Lutz (1985), let the time series comprising the signals be wrapped around a drum whose circumference is the periodicity under investigation. Asssociate a unit vector from the origin with each signal. If there is no periodicity the sum of these vectors will be a random walk; otherwise there will be some tendency for the vectors to occur in a particular sector and, depending on the strength of the non-randomness, their sum will depart more or less from that expected for a random walk. The probability distribution of a random walk in the plane, with unit step lengths, is known (Rayleigh 1880; Chandrasekhar 1943), and so the probability that the vector sum of the data is a departure from randomness can be assessed.

In this situation the null hypothesis is that the data are random, independent, and uniformly distributed. Departures from this simple null hypothesis may be caused not only by periodicity in the data but also by edge effects, long-term trends in the data (e.g. bunching into superchrons) or their interdependence (e.g. clustering). In situations where the data show long-term trends (a non-stationary time series) a modified random walk theory may still be applied, but in practice it is then easier to derive confidence limits by Monte Carlo methods, testing the real power spectrum against those generated from random data with the same underlying distribution. The confidence level to be assigned to a peak in the power spectrum depends on whether the corresponding periodicity is being abstracted from the data *ab initio* or whether it is a prediction of any particular hypothesis.

For a precisely defined period P_0, and with the above null hypothesis, a single trial confidence level may be obtained based on its estimated probability $p(I \geq I_0) = \exp[-0.5\ I(P_0)]$ of arising by chance. Here $I = 2R^2/N$ where R is the magnitude of the vector sum and N represents the number of independent data. This 'single trial' confidence level is then adjusted to the true confidence level by allowing for the equivalent number of trials involved in arriving at it. The latter is given by $n = \delta\nu/\Delta\nu$, where $\delta\nu$ represents the frequency interval corresponding to the range of periodicities being investigated, and $\Delta\nu = T^{-1}$ represents the fundamental frequency corresponding to the effective length T of the time series. The interval $\delta\nu$ being searched is the entire range of the data if periodicities are being sought *ab initio*, or the range within which a prescribed period may lie if some prior hypothesis is being tested. Thus, as has been stated, the confidence level one assigns to a derived periodicity depends on whether it may legitimately be regarded as a prediction of any particular theory.

The approach described above is easily modified to allow for variations in the quality of data within a given set (the vectors are given variable rather than unit length). The standard error of any derived periodicity may be derived by a maximum likelihood technique. In this approach one first finds the distribution of departures of the real data from the strict periodicity. Synthetic data are then generated which simulate both the derived periodicity and the 'error' distribution. Power spectra are applied to these synthetic data, and the distribution of synthetic periodicities yields the standard deviation of the real one.

In the numerous discussions on the topic published so far, periodicities have been extracted from the data without a prior theoretical framework. However I shall anticipate later discussion by stating that a 30 Myr cycle is reasonable expectation of the galactic theory, but that this cycle may be interspersed with a weaker 'interpulse' leading, in the best data, to an overall ~15 Myr one. The ~15 Myr cycle is a quarter of the Sun's period of oscillation perpendicular to the galactic plane. The presence of a real 15 Myr cycle in a wide range of good terrestrial data would thus strongly support the basic catastrophist thesis. Bahcall & Bahcall (1985) found that the half-period of the Sun's vertical motion out of the galactic plane may lie in the range 26-37 Myr, with maximum heights above the plane in the range 49-93 pc. The longer periods and heights were found in models wherein substantial unseen material is associated with a spheroid of half-height 700 pc, but Gilmore & Wyse (1987), from an analysis of K dwarfs at the south galactic pole, find no evidence for significant hidden mass beyond ~100 pc. A realistic range of solar half-periods and vertical excursions is then found to be 26-32 Myr, and 50-82 pc.

The Sun is close to the galactic plane, the time since its last passage through the plane being less than 3 Myr (Bahcall & Bahcall 1985). If Oort cloud disturbance of the long-period comet system is involved in terrestrial disturbances, the typical half-period of a long-period comet must be subtracted from the phase, implying that the Earth is currently in or almost in a bombardment episode. If on the other hand the primary source of disturbed comets is an unseen cloud of at least $\geq$3000 AU radius (galactic tides are inappreciable at smaller distances) then the infall time is ~0.15 Myr.

Thus for purposes of statistical testing, we may say that the galactic hypothesis predicts periodicities in the range 26-32 Myr and possibly 13-16 Myr, with phase 0-3 Myr. Some terrestrial data are now examined in the light of these expectations.

Impact cratering. It was first suggested by Seyfert & Sirkin (1979), on the basis of an inhomogeneous data set of mixed quality, that the terrestrial impact cratering record was periodic, with cycle length 26 Myr and phase such that the most recent 'impact epoch' began 2 Myr ago. An analysis of 11 impact craters with diameters $\geq$10 km and age uncertainties $\leq$20 Myr was carried out by Alvarez & Muller (1984), who obtained a very similar period of 28.4 $\pm$ 1 Myr, the slight difference being attributable to their arbitrary neglect of craters $\leq$ 5 Myr old (*cf.* also Trefil & Raup 1987). The result was regarded as significant at a confidence level ~99 %. Tremaine (1986) has however criticised the analysis on technical grounds, while Heisler & Tremaine (1989) have examined the effect of dating errors in the impact craters used by Alvarez & Muller (1984) and have drawn the conclusion that a periodicity even at 90% significance could not be obtained from the data, given the uncertainties in crater ages. They therefore consider that the apparent periodicity must be due to 'a statistical fluke or subjective bias'. Similar conclusions have been reached by Grieve *et al.* (1987).

These negative conclusions were, however, based on the *a posteriori* premise. If, on the other hand, prior knowledge of the astronomical environment is fed into the system, a different conclusion emerges. In order to test for a period as short as 13-16 Myr only the most accurate data may be employed; thus we select (Clube & Napier 1989) all known terrestrial impact craters greater than 5 km in diameter, younger than 250 Myr, and with formal dating uncertainties $\leq$5 Myr. These data have been culled from the lists of Grieve *et al.*, Shoemaker & Wolfe (1986) and Trefil & Raup (1987), and the Manson crater has been added (Hartung *et al.* 1989). Excluded from the list however are those considered to have been caused by the impact of iron bolides, probably derived from the main belt asteroids. A period of 16.2 $\pm$ 0.2 Myr is found. If the ~15 Myr period is taken to be a prior expectation of the galactic hypothesis, then the confidence level of the result is ~96%. The phase is found to be 2.9 $\pm$ 1.5 Myr (*cf.* 13 $\pm$ 2 Myr obtained by Alvarez & Muller), again as expected. With a chance expectancy ~20% for the phase agreement, the data seem to fit the prior hypothesis at a moderately high confidence level (~99%).

Geomagnetic reversals. The power spectrum technique discussed above cannot be applied unmodified to the geomagnetic reversal frequency data since contiguous events may not be independent (analogously to the clustering in age of some craters). If, however, the spurts in reversal frequency discussed by Pal & Creer (1986) may be regarded as independent, then the technique may be applied to their peaks. Not too much information is necessarily thrown away by this approach since a peak 'summarises' the behaviour of the field for some Myr before and after it. Problems with this approach are that the precise locating of a peak is liable to be a subjective or biassed activity, and to depend on the binning of the data. Further, the data for the last 250 Myr are not of equal quality, that earlier than 165 Myr being seriously incomplete (Pal 1988).

In an attempt to allow for these factors, the geomagnetic data to 165 Myr and 250 Myr (Harland *et al.* 1982) were each divided up into bins of various widths, the location of the peaks measured and power spectrum analyses applied. The results are illustrated in Figure 1 for bin widths 1 Myr and 2.5 Myr. There is in each case a peak at ~15 Myr, the formal significance of which ranges from dubious (97%) to extremely high (99.95%), for a single trial probability, when measured against the simple null hypothesis of random, uniform, independent data. However, the long-term behaviour of the record is clearly non-uniform, and allowing for this through simulations (*cf.* Lutz 1985), a signal at 15.3 ± 0.3 Myr is found, at an overall confidence level ~99%. Lutz & Watson (1988) have carried out simulations which revealed that ~28% of Fourier spectra of synthetic data randomly drawn from the long-term distribution had a highest peak at 30 ± 1 Myr; but an examination of their Figure 3 reveals that only ~1% of these simulations yielded a spurious peak at ~15 Myr: randomness in a structured setting like that of the observed reversal record may therefore yield a spurious 30 Myr cycle but not, with any degree of plausibility, a 15 Myr one.

It could be arbitrarily held that much of the 'confidence' of this result lies in a physically real 30 Myr cycle on which noise is superimposed, fortuitously exaggerating a 15 Myr harmonic of the basic cycle. Power spectra were therefore derived for data randomly extracted from a noisy 30 Myr sinusoidal variation extending over

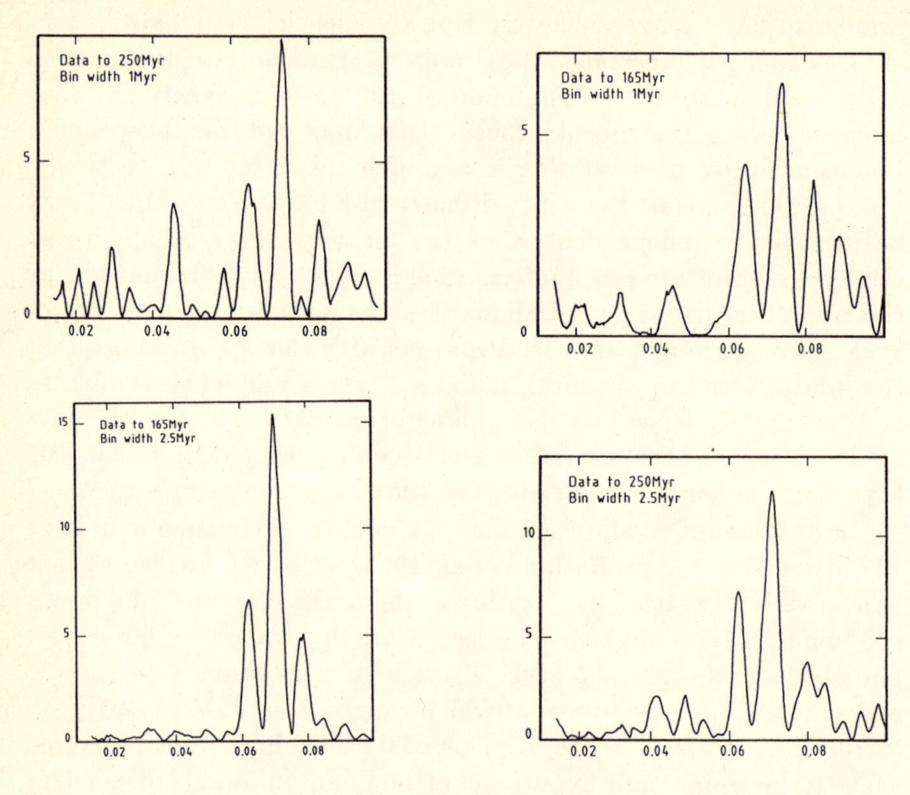

Figure 1. **Power spectra for peaks in geomagnetic reversal rate as determined for different bin widths. A peak appears at P∼15.5 Myr in all cases although with varying confidence levels (0.97≤C≤0.995).:**

150 Myr, the noise level being adjusted so that the basic cycle was recognised at a confidence level ≥ 99% in 50 out of 100 trials. In none of these trials was a spurious 15 Myr cycle detected at more than 90% confidence level.

The occurrence of a physically forced 15 Myr cycle, not an artefact of record length or non-stationarity and not a harmonic of a 30 Myr cycle, seems therefore to be established with ∼ 99% confidence for the geomagnetic reversal record, as expected for the galactic theory over a wide range of circumstances. This is consistent too with the fact that magnetic reversals have been associated with three of the five tektite age groups (Glass & Heezen 1967, Durrani & Khan 1971). If climatic modulation is seen as an intermediary between cometary disturbance and geomagnetic reversal (Hoyle 1982; Clube & Napier 1982a, 1986, 1989; Muller & Morris 1986), then it is also relevant that a relationship between reversals and climatic events seems to be well established at least for the upper Pleistocene (Doake 1978, Burek & Wanke 1988). On the other hand, the phase of the geomagnetic reversal cycle is 8.5 $\pm$ 0.7 Myr, out of match with the cratering record and significantly different from that expected from astronomical forcing. However, the reversal record of the last 90 Myr yields a period 16.1 $\pm$ 0.6 Myr, with phase 4.0 $\pm$ 2.2 Myr, which is consistent with both the cratering record and the astronomical expectations. It may therefore be that either real phase shifts, or simply inaccurate dating for the more distant past, are introducing a bias. Nevertheless the discrepancy remains to be resolved.

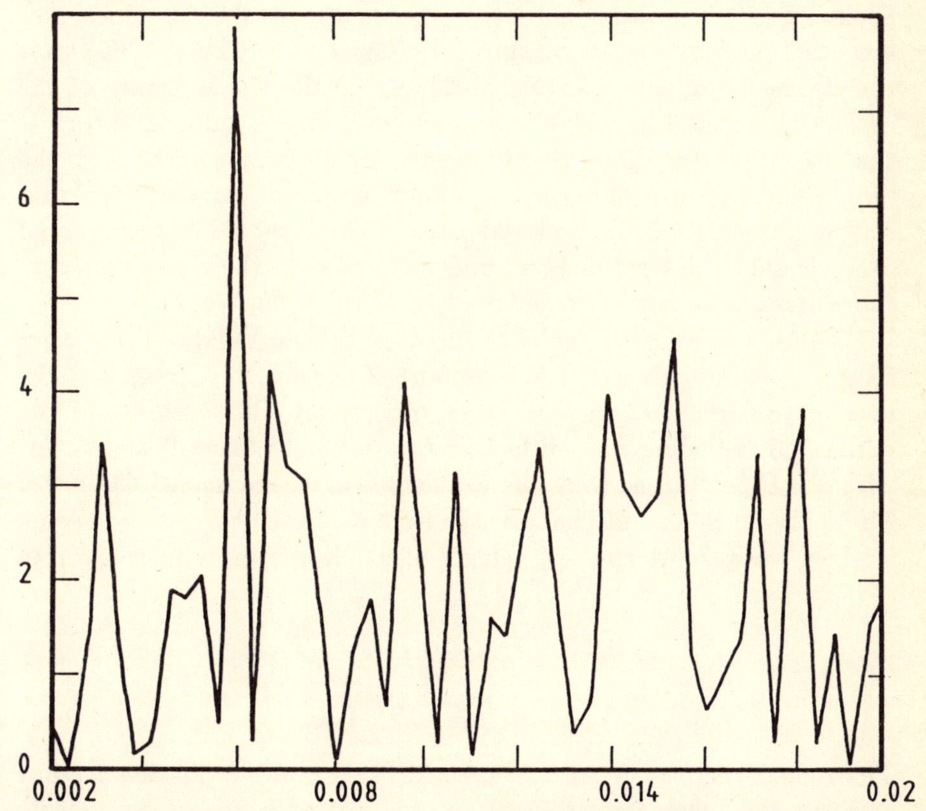

Figure 2. Power spectrum of epochs of worldwide vulcanism identified and dated by Pandey & Negi (1987). There is an overall fit to the expectations, significant at C∼0.98.:

Vulcanisms. Similar procedures have been applied to the compilation of worldwide vulcanisms given by Pandey & Negi (1987) (see Figure 2). In this case, however, one has to ask what constitutes a 'global vulcanism', and the epochs of these given by Pandey & Negi (which match well with those of Seyfert & Sirkin 1979) were uncritically adopted. A periodicity of $\sim$ 16 Myr was found, with phase 0-2 Myr. Given the data, the confidence level of the periodicity itself is only ~97% but coupled with the phase information the data seem to fit the galactic hypothesis at an overall confidence level ~99%.

Mass extinctions. The original evolution study by Raup & Sepkoski has been extended by them (1986) and a 26 Myr periodicity over the last 250 Myr, with confidence level ~99.9%, is now claimed. The various criticisms raised against these studies are discussed by Sepkoski (1989), who vigorously defends the work. He does not, however, address the criticism of Heisler & Tremaine (1989) that the claimed confidence level is too high for the r.m.s. errors of the extinction peaks ($\geq$ 6 Myr). A prediction of the galactic theory is that weaker 'interpulses' should eventually be found, perhaps in the 'insignificant' peaks of extinction culled from the analyses (Sepkoski & Raup 1986, Raup & Sepkoski 1986). These should have occurred over roughly the last 90 Myr, mid-way between the sharp peaks in the per-genus extinction rate found by Raup & Sepkoski.

These studies (see Table 2) indicate that a 15 Myr cycle does seem to emerge when the best extant data (possibly excepting the mass extinction data) are analysed objectively, this period and the derived phase being for the most part consistent (as will be shown) with the expectations from the astronomical environment. Claims of a ~30 Myr cycle seem to have arisen from a lack of objectivity in some studies, weaker 'interpulses' being neglected, or simply from paucity of data.

Nemesis

The rediscovery, by Raup & Sepkoski (1984), of the Fischer-Arthur cycle in the marine microfossil record, not only stimulated a cottage industry in period-hunting, it led to a search for explanations. Rampino & Stothers (1984) proposed that the Oort cloud was being

Table 2. Periodicities derives from several terrestrial phenomena, compared with expectations from galactic theory.:

phenomenon	cycle length (Myr)	phase (Myr BP)
galactic cycles (predicted)	29.0±3.0 14.5±1.5	1.5±1.5
impact craters	16.2±0.2	2.9±1.5
geomagnetic reversals (to 165 Myr)	15.3±0.3	8.4±0.7
geomagnetic reversals (to 90 Myr)	16.1±0.6	4.0±2.2
global vulcanisms	~16	1±1

disturbed by interactions with individual molecular clouds in the course of its vertical galactic oscillations, while two groups were led to propose that the Sun had a companion star with an orbital period of 26 Myr. The latter proposal is discussed here.

The companion star, dipping into the Oort cloud every 26 Myr, perturbs comets into orbits taking them into the planetary system, whence bombardment episodes recur periodically. The Whitmire & Jackson (1984) version of the hypothesis has the companion in a highly eccentric orbit ($e \geq 0.8$), whereas in the Davis, Hut & Muller (1984) variant the eccentricity is variable, fluctuating around its phase-averaged value $e \sim 0.7$. For the binary star hypothesis to work, it is necessary to supplement it with a further hypothesis, namely that the Oort cloud possesses an unseen dense inner core. The status of this supplementary hypothesis is discussed by Bailey, Clube & Napier (1989). The binary star hypothesis thus also adopts Oort cloud disturbance and bombardment episode as the basic mechanism for controlling Earth processes, but because the perturber is now a star the bombardment episodes may be sharper and more intense than those of the galactic model.

There are three immediate scientific objections to the Nemesis proposal. In the first place the system postulated is of a kind which is, so far as is known from many sky surveys, non-existent (Clube &d Napier 1984b). Common proper motion pairs, which comprise the widest known binaries, have a computed mean period 67,000 yr and median 3,100 yr. It is true that filters applied to observed binaries will tend to keep out physical wide binaries as well as mere optical doubles from the catalogues, but this may be circumvented by examining catalogued multiple systems to which the filters have not been applied. In this way Abt (1986, 1988) finds that amongst stars of solar type and age, an upper limit to the separation of $\sim$ 2,000 AU is expected. This should be compared with the postulated semi-major axis of Nemesis, $\sim$ 92,000 AU, and its aphelion distance $\sim$ 150,000 AU for its mean eccentricity.

Secondly, the orbit is unstable (Table 3). Even on the most conservative assumptions, the energy injected into a binary of this separation through encounters with stars and molecular clouds is, over the age of the Solar System, $\sim$200 times the binding energy of the companion to the Sun. A formal survival time of 50-100 Myr was estimated by Clube & Napier (1984), although in reality the companion might survive until its first encounter with a giant molecular cloud: there may have been 5-15 such encounters over the lifetime of the Solar System. (Tremaine (1986) estimated a formal half-life of about 400 Myr, but this is much too long, being based on a formula by Hut & Tremaine (1985) which greatly underestimates the effects of molecular clouds: see Clube & Napier (1986), Bailey (1986), Bailey

Table 3. Formal half-lives (in Myr) of Nemesis for various assumed initial semi-major axes (in AU), perturbed by stars or molecular clouds. From formulae by Bailey (1986):

Semi-major axis (AU)	20000	30000	90000
stars	4000	2700	900
molecular clouds	220	65	2.4

et al. (1989). Hut (1984) estimated a half-life $\sim$1000 Myr starting from the present orbit by ignoring the major perturbers, molecular clouds, altogether). One might hope to protect the companion star by having it in a more tightly bound orbit which has recently, in geological terms, been prised loose and is temporarily yielding a 30 Myr cycle: we would then by chance be 'detecting' the star during the last few 100 Myr of its 4500 Myr of captivity. However a tighter orbit implies a much higher comet flux into the planetary system and a much higher cratering rate (several powers of ten, depending *inter alia* on the assumed structure of the Oort cloud). In fact the lunar cratering rate has been remarkably constant for the last 3.9 Gyr, a result which is inconsistent with the supposed binary history.

Finally, a companion star should reveal itself in the current Oort cloud. Any surviving cloud would betray gross asymmetries corresponding to the injection of angular momentum in directions perpendicular to the recent plane of motion of the star. As will be discussed, the comets of the Oort cloud reveal the effects of the galactic tide and of a recent encounter with (probably) a nearby nebula. Of a companion star, however, there is no sign. The hypothesis seems to have been motivated by a misconception, namely that the galactic mechanism was incapable of producing a 30 Myr cycle. This, as we shall see, is incorrect.

Galactic structure and terrestrial cycles

The Oort cloud is acted on by both random and structured forces: the random component is dominated by encounters with stars and giant molecular clouds, the systemic one by the disk and spiral arms of the Galaxy. Knowledge of these various perturbers allows one to predict theoretically the importance of, say, short-lived comet showers of order a few Myr against, say, more extended episodic variations on timescales of order 100 Myr. Conversely, the geological record has the potential to constrain uncertain properties of the galactic environment.

Stars. It is now generally recognised that random encounters with stars and molecular clouds have the potential to produce 'bombardment episodes' or 'impact epochs' on the Earth, the duration and intensity of which depend on the unknown structure of the Oort cloud. Stars penetrate the Oort cloud at a mean rate of $\sim$ 1 per Myr, and on the assumption that there is a dense inner cloud of comets, brief comet blizzards have from time to time been postulated to account for mass extinctions. Hills (1981) argued that during an intense comet shower a comet flux as high as one per hour might be anticipated, and he proposed this as an explanation of the extinctions of 65 Myr ago.

If the energy distribution of Oort cloud comets is taken to be $n(E) \propto |E|^{-\gamma}$, the radial density distribution is roughly a power law with index $\gamma - 4$. The classical Oort cloud has γ=2.5, massive inner cloud models generally having negative γ. Over 100 Myr there is an expectation that one star of mass $\geq 0.6\ M_\odot$ will pass within 7800 AU of the Sun. From Fernandez & Ip (1987) one then finds that, say for a comet cloud with $\gamma = -2$, the number of comets thrown into the planetary system by such a passage would be enhanced by a factor ≥ 30 over the background rate. The cratering record would reveal strong clustering of ages at one or two epochs, contrary to the observed record. Such considerations have been used by Napier (1987), Stothers (1988) and Bailey & Stagg (1988) to argue that at least the extreme Hills-type models of the Oort cloud are improbable and that comet showers more than say an order of magnitude above background are not expected. Comet showers induced by stellar perturbations would not in any case yield periodicities in the various terrestrial records. On current evidence, then, it seems that intense comet showers are unlikely to have been major factors guiding terrestrial processes.

Molecular clouds. These are now generally recognised as the chief energisers of the Oort cloud. The question of Oort cloud stability in the presence of these massive perturbers has been intensively discussed, with values as high as 20000 Myr and as low as 700 Myr being quoted for the half-lives of the long-period comet system. The shorter lifetimes derived in earlier work have however now been confirmed in recent studies. Thus Bailey (1986) finds the half-life of a comet of semi-major axis a_4 (in units of 10^4 AU) to be $\sim 8a_4^{-1}$ Gyr due to stellar perturbations, and $\sim 1.8a_4^{-3}$ Gyr for molecular cloud perturbations. Over the lifetime of the Solar System, therefore, an Oort cloud with the classical structure would have been swept away several times over (Napier & Clube 1979, Clube & Napier 1982b, Napier & Staniucha 1982). Replenishment might happen in principle either by unbinding from a dense inner cloud or by capture from pre-existing comets in a molecular cloud (Napier & Clube 1979). This is a live issue at present (see Bailey *et al.* 1989), and it is recognised that the constraints on any dense inner cloud which may exist are tight (Napier 1987; Bailey & Stagg 1988).

From the point of view of terrestrial control, probably the main issue is whether encounters with molecular clouds would produce galactic periodicities. Rampino & Stothers (1984) have argued that the Solar System's motion perpendicular to the galactic plane is sufficient to modulate the encounter rate with molecular clouds and so lead to high and low comet fluxes with a 30 Myr cycle. This requires the half maximum $z_{1/2}$ of the medium-sized molecular clouds to be ≤ 60 pc. Thaddeus & Chanan (1985), estimating $z_{1/2} \sim 85$ pc, argued that no appreciable galactic modulation by molecular cloud encounters is possible (see also Bailey *et al.* 1987). More recent half-widths for molecular clouds and young stars (related to molecular clouds) show $z_{1/2} \sim 60$ pc, however, which, it is argued by Stothers (1988), would permit detection of a $\sim$ 30 Myr oscillation.

Numerical simulations carried out by the author reveal that a 30 Myr modulation of the encounter rate would indeed be just detectable with the parameters adopted by Stothers; however a 15 Myr cycle would not, being lost in the 'noise' of random nebular encounters. It appears, however, that the question of modulation by this process is probably a non-issue: it turns out that the systemic potential of the Galaxy is a more regular and more powerful modulator of the comet flux.

Spiral arms. The importance of the galactic tide in applying a steady perturbation to the orbits of long-period comets was appreciated by several groups almost simultaneously (Byl 1983, 1986; Muller & Morris 1986; Heisler & Tremaine 1986). It is readily shown that the tidal component perpendicular to the galactic plane is the dominant one, the radial component having only 5-20% percent of its strength. Adopting a plane-parallel approximation for the distribution of matter in the solar neighbourhood, and applying classical potential and perturbation theory (*e.g.* Byl 1986), it is found that the perihelion distance q and galactic latitude b of a comet vary cyclically. A relatively large change in q ($\Delta q \sim 50$ AU for $q=100$ AU) may occur from one revolution to the next, with evolution to very small q-values occurring, for an individual comet, periodically on timescales of a few 100 Myr (Byl 1986, Heisler & Tremaine 1986). Thus the existence of the tide ensures that, when comets enter the planetary system and are destroyed or ejected from the Solar System, there is a replenishing supply. The magnitude of the effect depends on whether there is an appreciable missing mass at low galactic heights.

Under the plane-parallel assumption, the strength of the tide (and hence the flux of comets into the planetary system) is directly proportional to the density of the ambient material. It follows that, if the density of galactic disc material varies appreciably with height above the plane, the comet flux will be modulated, and for modest amplitudes of solar oscillation a $\sim$ 30 Myr cycle will be induced. Unfortunately $z_{1/2}$ is not well known for low z values, and whether a cyclic variation in tide would occur at a detectable level, due to variations in ambient disc density, cannot be securely inferred from the astronomical observations. In any case, such a variation would not yield a 15 Myr cycle. The assumption of plane-parallelism may, however, be questioned. In Figure 3 is shown the spiral galaxy NGC 2997. This is a late- type, Sc spiral (in which the arms predominate), and is probably typical of its genre. Considerable fine structure is evident within the arms; in particular long, narrow dust lanes can be seen typically $\sim$1 kpc in length and $\sim$100 pc wide. These may be identified with the molecular cloud complexes discussed by Elmegreen (1987) and others. These contain 10^7 $M_\odot$ of molecular and atomic gas, and have been mapped locally in the Carina spiral arm (Grabelsky *et al.* 1987) and in the Sagittarius and Scutum arms (Elmegreen & Elmegreen 1987). There may be $\sim$20-100 cloud complexes in a typical spiral galaxy, comprising about half the total mass of the gas (Elmegreen 1987). The dimensions and masses of these complexes reveal that they may have large overdensities, of order 5 or 10, relative to the background of disc stars. Likewise both theory (*loc. cit.*) and 21 cm observations of other spiral galaxies (Elmegreen & Elmegreen 1984) indicate that spiral arms may have arm/interarm density contrasts of 4:1 or larger. It follows that strong departures from a plane parallel galactic tide are expected whenever the Sun, in its galactic orbit, passes near a dust lane or spiral arm. According to Urasin (1987), for example, the Sun is near the inner edge of the Orion spiral arm, adjacent to a spur extending from it. Thus the potential in the solar neighbourhood may not even remotely be plane-parallel, a fact which should manifest itself both in local stellar kinematics (Woolley 1965) and in tidal action on the Oort cloud.

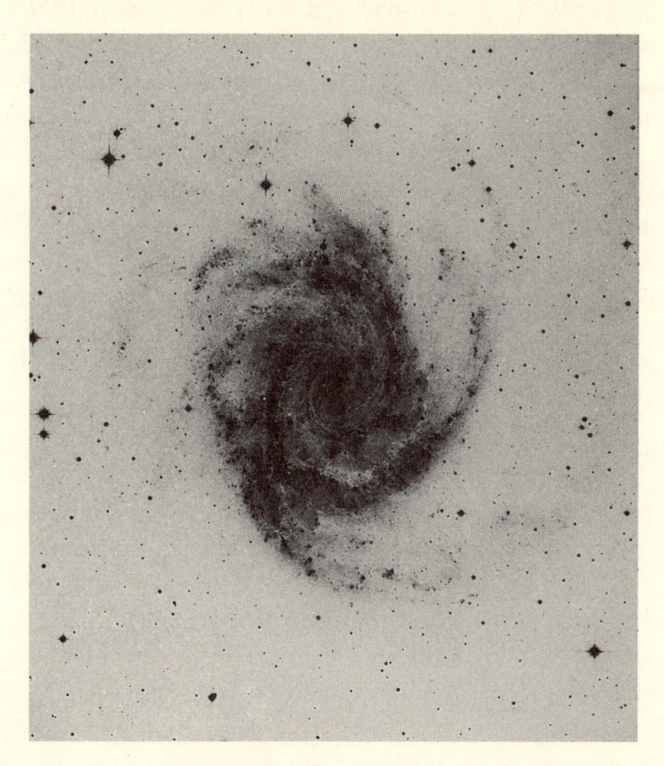

Figure 3. **The spiral galaxy NGC 2997, revealing considerable fine structure, including numerous narrow dust lanes (white streaks in this negative) imbedded in broader spiral arms. (Courtesy the Observatories of the Carnegie Institution):**

The solar direction of motion is such that the Sun has recently emerged from the Orion spiral arm. In general, however, the crossing time through the arm is a function of pitch angle, peculiar solar motion and arm motion relative to the stellar background. The pitch angle of the local arm is ~7°(Pavlovskaya & Suchkov 1980, Urasin 1987), and the peculiar solar motion is ~ 13 km s^{-1}. The pattern speed of the spiral arms in the solar neighbourhood appears to be in the range 20-25 km s^{-1} kpc^{-1} according to a number of workers (Stromgren 1967, Creze & Menneseer 1973, Talbot 1980, Marochnik 1983 *etc.*). This is substantially higher than expected on the Lin-Shu density wave theory and may be interpreted to mean either that the theory needs to be modified (Marochnik 1983), or that some quite different one is appropriate (*e.g.* Ambartsumian 1965, Clube 1978). At any rate, the time taken for the Sun to cross the Orion spiral arm was, according to these figures, within 50% of 50 Myr. Thus one envisages the Sun crossing the Orion spiral at a fairly narrow angle to the axis, completing one or two porpoise-like cycles as it does so.

Variations in the tide exerted on the Oort cloud (and hence comet flux in the planetary system) may thus be calculated for various galactic parameters and positions of the Sun. For this purpose, a spiral arm may be represented as a long circular or elliptic cylinder imbedded in a galactic disc with exponential density distribution (see Clube & Napier 1989 for details). The uncertain scale height of the disc material turns out to be irrelevant and the disc can be represented as a uniform slab for the purpose. In the extreme case where the disc is neglected altogether, powerful surges, recurring at 30 Myr intervals, are evident. These are readily understood as the Sun, successively emerging from and falling into the spiral arm, finds itself alternately in regions of high and low ambient density. For the more realistic case of moderate spiral arm (or dust lane) overdensities, strong 30 Myr surges are still evident but there are now weaker 'interpulses' which may yield, in principle, a 15 Myr cycle with weak and strong cycles interpersing.

The interpulses arise because, above a spiral arm, there are two opposing tidal components. The tide due to the ambient disc material is compressive, that due to the spiral arm is tensile. These tend to cancel close to the arm, but as the Sun rises towards the peak of its orbit (say 80 pc above the plane) the tensile force of the arm declines, and the compressive one of the disc begins to dominate, so increasing the overall stress. This secondary effect seems to be appreciable for arm/interarm contrasts in the range 2:1 to 4:1, but declines as more extreme density contrasts are encountered. Given all the uncertainties, it is nevertheless satisfactory that a strong 30 Myr cycle, presumably recognised as a 15 Myr one in the most complete geological records of the last 50-100 Myr, is a reasonable expectation of the observed galactic environment.

To summarise, the chief objection frequently raised against a galactic modulation of the terrestrial impact rate (Thaddeus & Chanan 1986, Tremaine 1986, Bailey *et al.* 1987 *etc.*), is that the scale height of the molecular cloud system is too great for a 30 Myr cycle to become evident. This statement is disputed by Stothers (1988), but the issue seems to be irrelevant as it refers only to the stochastic component of the galactic potential field. Following the realisation that galactic tides matter, it appears that strong driving of a weak-strong 15 Myr cycle, probably resolved as a 30 Myr one in a very incomplete geological record, is caused by the overall, systemic potential structure of the spiral arms and the disc. These cycles will persist for as long as the Solar System is in the neighbourhood of a spiral arm.

A corollary is that, between arms, such modulation will decline. If the scale height of the disc material is comparable to the amplitude of the solar oscillation, a 30 Myr cycle will persist at some level but it is a prediction of the hypothesis that no 15 Myr one will be evident. To the extent that terrestrial phenomena are affected by the cometary flux, therefore, they are expected to be episodic on characteristic time scales of around 50-100 Myr.

The current neighbourhood

A yet more detailed look at the immediate galactic environment reveals that the Solar System, on emerging from the Orion spiral arm, penetrated what appears to be the remains of an old, disintegrating giant molecular cloud. This is now a ring of material (Lindblad's Ring), expanding from some event of 30 Myr ago, which incorporates most of the molecular clouds and star-forming regions in the solar neighbourhood. The young blue stars form an arc in the sky now known as Gould's Belt, but recognised since the time of Ptolemy. The solar antapex coincides with the convergent point of the Scorpio-Centaurus association (Figure 4), implying that the Solar System not only passed through Gould's Belt, but also through or at least very close to this specific sub-structure. The encounter seems to have taken place only 5-10 Myr ago (Napier & Clube 1979). Evidence of this recent encounter should manifest itself not only in recent terrestrial disturbances but also in an unrelaxed Oort cloud.

Ever since the Oort cloud concept was formulated in a quantitative way (Oort 1950), a problem has been the observed concentration of semi-major axes towards large distances (near zero energy). The probem arises because mechanisms based on for example formation and ejection of comets from the planetary region yield a wider energy distribution than is observed. This difficulty is 'overcome' by postulating a fading function, by which comets at smaller semi-major axes, having previously passed through the planetary system, are rendered inactive with some degree of probability. There is no independent evidence for a fading function of the required specific form, however (Bailey 1984), and it remains an *ad hoc* device whose function is to remove a contradiction between theory and observation. Yabushita (1979) has argued that the observed energy distribution of comets is on the other hand consistent with a capture of comets, the event having taken place 5-9 Myr ago. Within the errors this seems to be consistent with the recent encounter with a molecular cloud (Clube & Napier 1984).

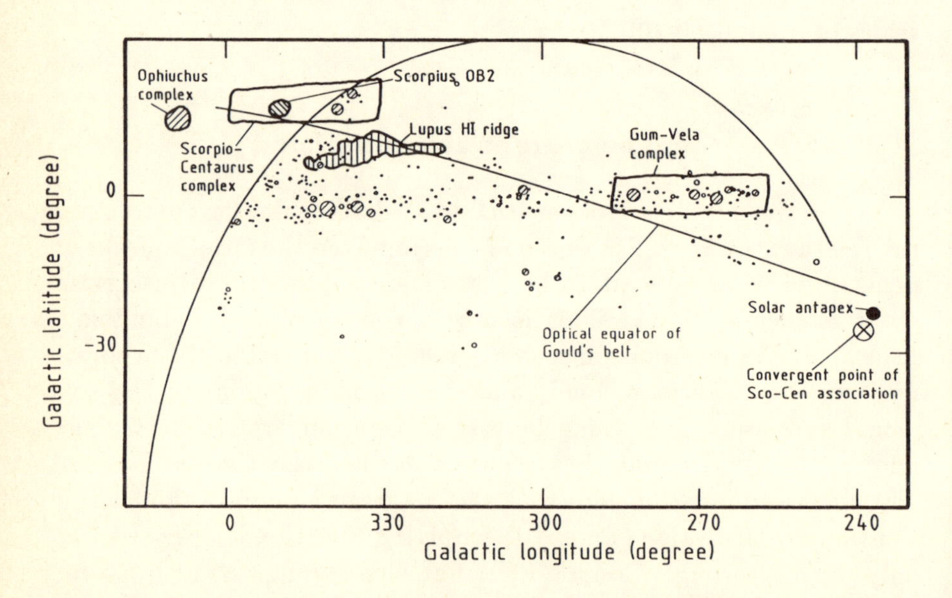

Figure 4. Gould's Belt from the Earth, from a recent catalogue of nebulae. Note that the solar antapex and the convergent point of the stars of the Scorpio-Centaurus association are virtually coincident, again indicating that recent terrestrial and comet cloud disturbances are expected on the galactic theory.:

Recent studies by Delsemme (1987) have confirmed earlier work that there is a dearth of long-period comets with aphelia around the galactic poles and equator. This is as expected if the arrival of comets is controlled by the galactic tides, but Yabushita (1989) has shown that the orientation parameters of these comets are not in fact consistent with an equilibrium tidal driving and proposes instead that either there have been exceptional stellar or nebular perturbations of the comet system during the past 5-10 Myr, or the comets have been recently captured and have not yet had time to settle into an equilibrium distribution dictated by galactic tides.

Delsemme's investigation also revealed, in addition to this broad structure of comet aphelia over the celestial sphere, a distinctive stream of 36 comets with a general retrograde revolution lying in a narrow strip which contains both the galactic centre and the solar antapex (and thus the Sco-Cen convergent point). But because the solar apex moves over the sky at a rate of $\sim 1^{\circ}.5$ Myr^{-1}, this phenomenon, if due to a discrete event, must have been caused by a very recent disturbance of the Oort cloud related in some way to the solar motion. It may be shown that a recent stellar encounter is not a viable explanation for the phenomenon, and it seems most likely that this comet strip arose from a disturbance caused by substructure within the molecular cloud which gave birth to the stars of the Scorpio-Centaurus association. It seems then that the comets which are currently arriving into the planetary system are not part of a relaxed system but rather show signs of recent disturbance (Clube & Napier 1984). This is consistent with expectations from the immediate galactic environment.

Whether one should regard the Gould Belt disturbance as part of the systemic potential or the stochastic one is perhaps a question of semantics rather than physics: the Gould Belt material is both a discrete system and part of the Orion spiral arm. At any rate it is satisfactory that a detailed analysis of the current environment leads to a general concordance with the broad theme of galactic control.

Remarks in conclusion

The basic thesis is that the galactic environment exerts a close control over major physical and evolutionary processes on Earth. Since this theory was first introduced (Clube 1978, Napier & Clube 1979), new astronomical data have revealed the importance of very large comets and galactic tides, and it may be that these discoveries have completed the causative chain connecting Earth and Galaxy. The chain, nevertheless, still possesses weak links. One of these is that detailed mechanics have yet to be worked out for even the major terrestrial processes likely to be at work during a disturbed epoch, and the complexity and non-linearity of the Earth machine are such that this situation is likely to pertain for a long time. Indeed it is not even clear what the major processes are. Figuratively speaking, there are too many ways of killing dinosaurs even within the context of the theory. Consider the proposition '10 km diameter bolide = dust in stratosphere = collapse of food chains = KT extinctions'. Although this series of equations was originally proposed only as a contributory factor, it has since become the mainstay of the 'giant impact' school. However the optical depth of stratospheric dust veils saturates at bolide masses $\geq 10^{16}$ gm corresponding to impacts at $\sim$ 1 Myr intervals (Clube & Napier 1982a), a statement which seems to be supported by the detailed calculations of Ahrens & O'Keefe (1982) for the impact of a 10 km bolide. These prompt dustings are much too frequent to be the prime cause of great mass extinctions. At present it seems more plausible that, if a single giant impact did indeed cause sudden death at the end of the Mesozoic, the blast wave did it. But the theory lacks definiteness at this point. Thus the effects of a bombardment episode are likely to be multifarious and complex (Clube & Napier 1982a), and although there has been much detailed modelling of the prompt effects of a single giant impact, it is not obvious that any of it bears much relation to the major factors likely to be at work during a bombardment episode of few million years duration.

Indeed, many of the objections frequently raised against terrestrial catastrophism (*e.g.* sea-level regressions correlate with extinction episodes, specific extinctions do not precisely coincide with adjacent microtektite layers etc.) apply only to the simple 'giant impact' model of catastrophism: the dominance of this concept in contemporary thinking on the subject has led to the frequent pursuit of red herrings. On the other hand, while it is true that a more mature understanding of the astronomical environment removes these false dichotomies, it is also true that the range of possible trauma is, in the present state of knowledge of that environment, almost too wide to be useful for prediction.

Another, more serious, uncertainty lies in the assumption of a direct link between galactic perturbations and comet influx to the inner planetary system. The rate at which short-period comets currently arrive in the planetary system is one or two powers of ten higher than the steady-state rate of supply from the long-period comet system. This may reflect an imbalance due to a recent disturbance, or even the recent disintegration of a giant comet, but it has also been proposed that the source of the short-period comets is not the classical Oort cloud but rather a dense, unseen inner cloud. Galactic perturbations probably reach down to $\sim$ 3,000 AU, but it has been suggested that the hypothetical source is a ring just beyond the planets (Duncan *et al.* 1988; but see Stagg & Bailey 1989). Should future work reveal that the short-period comets derive largely from such a ring, then a galactic modulation would be difficult to sustain (as would any 'comet shower' proposal) and other explanations would have to be sought for the 'galactic' cycles. While the postulated link is probable, it is not yet proven.

Despite these uncertainties, the galactic theory seems to have some advantages. Parsimony is one: it is closely based on the specifics of the observed astronomical environment, without *ad hoc* additions. Predictive power is another: the likely discovery of 15 and 30 Myr cycles in the terrestrial record is a particularly encouraging development. The shorter period should be detectable, in principle, in the extinction record, and a quantitative demonstration of 15 or 30 Myr cycles in sea-level variation would further strengthen the theory.

The idea of disturbing the Oort cloud by a companion star with a 30 Myr period seems on the other hand to meet with insuperable difficulties and is in any case superfluous. However, outrageous hypotheses do sometimes have value (Feyerabend 1975), and the Nemesis proposition has certainly stimulated further work on the more realistic problem of Oort cloud evolution in a galactic setting.

References

Abt, H.A., 1986 *The ages and dimensions of Trapezium systems.* Astrophys.J. **304**, 688.

Abt, H.A., 1988 *Visual binary separations as functions of primary types, ages and locations.* Astrophys. Space Sci. **142**, 111.

Ager, D., 1975 *Major marine cycles in the Mesozoic.* J Geol. Soc. **138**, 159.

Alvarez, W., Alvarez, L.W., Asaro, F. & Michel, H.V., 1980 *Extraterrestrial cause for the Cretaceous-Tertiary extinction: Experimental results and theoretical interpretation.* Science **208**, 1095.

Alvarez, W. & Muller, R.A., 1984 *Evidence from crater ages for periodic impacts on the Earth.* Nature **308**, 718.

Bahcall, J.N. & Bahcall, S., 1985 *The Sun's motion perpendicular to the galactic plane.* Nature **316**, 706.

Bailey, M.E., 1984 *The steady-state 1/a distribution and the problem of cometary fading.* Mon.Not.R.astr.Soc. **211**, 347.

Bailey, M.E., 1986 *The mean energy transfer rate to comets in the Oort cloud and implications for cometary origins.* Mon.Not.R.astr.Soc. **218**, 1.

Bailey, M.E., Clube, S.V.M. & Napier W.M., 1990 THE ORIGIN OF COMETS Pergamon, Oxford.

Bailey, M.E. & Stagg, C.R., 1988 *Cratering constraints on the inner Oort cloud: steady-state models.* Mon.Not.R.astr.Soc. **235**, 1.

Bailey, M.E., Wilkinson, D.A. & Wolfendale, A.W., 1987 *Can episodic comet showers explain the 30 Myr cyclicity in the terrestrial record?* Mon.Not.R.astr.Soc. **227**, 863.

Baldwin, R.B., 1985 *Relative and absolute ages of individual craters and the rate of infalls on the Moon in the post-Imbrium period.* Icarus **61**, 63.

Burek, P.J. & Wanke, H., 1988 *Impacts and glacio-eustasy, plate-tectonic episodes, geomagnetic reversals: a concept to facilitate detection of impact events.* Phys. Earth Planet. Interiors **50**, 183.

Byl, J., 1983 *Galactic perturbations of nearly-parabolic cometary orbits.* Moon and Planets **29**, 121.

Byl, J., 1986 *The effect of the Galaxy on cometary orbits.* Earth, Moon, Planets **36**, 263.

Chandrasekhar, S., 1943 *Stochastic problems in physics and astronomy.* Rev. Mod. Phys. **15**, 1.

Clube, S.V.M., 1978 *Does our Galaxy have a violent history?* Vistas in Astron. **22**, 77.

Clube, S.V.M. & Napier, W.M. 1982a *The role of episodic bombardment in geophysics.* Earth Planet. Sci. Lett. **57**, 251.

Clube, S.V.M. & Napier, W.M. 1982b *Spiral arms, comets and terrestrial catastrophism.* Q.J. Roy. astr. Soc. **23**, 45.

Clube, S.V.M. & Napier, W.M. 1984a *Comet capture from molecular clouds: a dynamical constraint on star and planet formation.* Mon.Not.R.astr.Soc. **208**, 575.

Clube, S.V.M. & Napier, W.M. 1984b *The microstructure of terrestrial catastrophism.* Mon.Not.R.astr.Soc. **211**, 953.

Clube, S.V.M. & Napier, W.M. 1984c *Terrestrial catastrophism: Nemesis or Galaxy?.* Nature **311**, 653 (see also **313**, 503).

Clube, S.V.M. & Napier, W.M. 1986 *Giant comets and the Galaxy: implications of the terrestrial record.* In THE GALAXY AND THE SOLAR SYSTEM (eds. R. Smoluchowski, J.N. Bahcall & M.S. Matthews), p.260. University of Arizona Press, Tucson, Arizona.

Clube, S.V.M. & Napier, W.M. 1989 *The 15 Myr galacto-terrestrial cycle.* In preparation.

Creze, M. & Menneseer, M.O., 1973 *An attempt to interpret the mean properties of the velocity field of young stars in terms of Lin's theory of spiral waves.* Astron. Astrophys. **27**, 281.

Davis, M., Hut, P. & Muller, R.A., 1984 *Extinction of species by periodic comet showers.* Nature **308**, 715.

de Laubenfels, M.W., 1956 *Dinosaur extinction: one more hypothesis.* J. Paleont. **30**, 207.

Delsemme, A.H., 1987 *Galactic tides affect the Oort cloud: an observational confirmation.* Astron. Astrophys. **187**, 913.

Doake, C.S.M., 1978 *Climatic change and geomagnetic field reversals: a statistical correlation.* Earth Planet. Sci. Lett. **38**, 313.

Duncan, M., Quinn, T. & Tremaine, S.D., 1988 *The origin of short-period comets.* Astrophy. J. **328**, L69.

Durrani, S.A. & Khan, H.A., 1971 *Ivory Coast microtektites: fission track age and geomagnetic reversals.* Nature **232**, 320.

Elmegreen, B.G., 1987 *Formation and evolution of the largest cloud complexes in spiral galaxies.* In PHYSICAL PROCESSES IN INTERSTELLAR CLOUDS (eds. G.E. Morfill and M. Scholer), p.1. D. Reidel, Dordrecht, Holland.

Elmegreen, B.G. & Elmegreen, D.M. 1987 *HI superclouds in the inner Galaxy.* Astrpophys. J. **320**, 182.

Feyerabend, P.K. 1975 AGAINST METHOD. Verso, London.

Fernandez, J.A. & Ip, W.H., 1987 *Time dependent injection of Oort cloud comets into Earth-crossing orbits.* Icarus **70**.

Fischer, A.G. & Arthur, M.A., 1977 *Secular variations in the pelagic realm.* Soc. Econ. Paleont. Mineral. Spec. Publ. **25**, 19.

Gallant, R.L., 1964. BOMBARDED EARTH John Baker, London.

Gilmore, G. & Wyse, R.F.G., 1987. *The multivariate stellar distribution function.* in THE GALAXY (eds. G. Gilmore and R. Carswell), D. Reidel, Dordrecht, Holland p.247.

Glass, B.P. & Heezen, B.C., 1967 *Tektites and geomagnetic reversals.* Sci. Am. **217**, 32.

Grabelsky, D.A., Cohen, R.S., Bronfman, L. et al., 1986 *Molecular clouds in the Carina arm : large scale properties of molecular gas and comparison with HI.* Astrophys. J. **315**, 122.

Grieve, R.A.F., Sharpton, V.L., Goodacre, A.K. & Garvin, J.B. 1985 *A perspective on the evidence for periodic cometary impacts on Earth.* Earth Planet. Sci. Lett. **76**, 1.

Hallam, A., 1984 *Asteroids and extinction — no cause for concern.* New Scientist 8 Nov., p 30.

Harland, W.B., Cox, A.V., Llewellyn, P.G. et al., 1982 A GEOLOGIC TIME SCALE Cambridge University Press, Cambridge.

Hartung, J., Kunk, M.J. & Anderson, R.R. 1989 *The Manson impact structure, a possible site for the Cretaceous-Tertiary (KT) boundary impact.* In Global Catastrophes in Earth History (Snowbird, Utah), p. 70. Contr No. 673, Lunar and Planetary Science Institute.

Heisler, J.S. & Tremaine, S. 1986 *The influence of the Galactic tidal field on the Oort cloud.* Icarus **65**, 13.

Heisler, J. & Tremaine, S., 1989 *How dating uncertainties affect the detection of periodicity in extinctions and craters.* Icarus **77**, 213.

Hills, J.G., 1981 *Comet showers and the steady-state infall of comets from the Oort cloud.* Astron. J. **86**, 1730.

Holmes, A., 1927 THE AGE OF THE EARTH: AN INTRODUCTION TO GEOLOGICAL IDEAS. Benn, London.

Hoyle, F., 1982 ICE. Dent, London.

Hoyle, F. & Lyttleton, R.A., 1938 *The effect of interstellar matter on climatic variation.* Proc. Camb. Phil. Soc. Math. Phys. Sci. **35**, 405.

Hoyle, F. & Wickramasinghe, C. 1978 *Comets, ice ages and ecological catastrophes.* Astrophys Space Sci **53**, 523.

Hsu, K.J., 1980 *Terrestrial catastrophe caused by cometary impact at the end of Cretaceous.* Nature **285**, 201.

Hut, P., 1984 *How stable is an astronomical clock that can trigger mass extinctions on Earth?.* Nature **311**, 638.

Hut, P. and Tremaine, S., 1985 *Have interstellar clouds disrupted the Oort cloud?.* Astron. J. **90**, 1548.

Innanan, K.A., Patrick, A.T. and Duley, W.W., 1978 *The interaction of the spiral density wave and the Sun's galactic orbit.* Astrophys. Space. Sci. **57**, 511.

Lindsay, J.F. & Srnka. L.J. 1975 *Galactic dust lanes and lunar soil* Nature **257**, 776.

Lowrie, W. & Alvarez, W., 1981 *100 million years of geomagnetic polarity history.* Geology **9**, 392

Lutz, T.M., 1985 *The magnetic reversal is not periodic.* Nature **317**, 404.

Lutz, T.M. & Watson, G.S., 1988 *Effects of long-term variation on the frequency spectrum of the geomagnetic reversal record.* Nature **334**, 240.

Macintyre, R.M. 1971 *Periodicty of carbonatite emplacement.* Nature Phys. Sci. **230**, 79.

Mardia, K.V., 1972 *Statistics of Directional Data.* Academic Press, London and New York.

Marochnik, L.S., 1983 *On the origin of the solar system and the exceptional position of the Sun in the Galaxy.* Astrophys. Space Sci. **89**, 61.

Mazaud, A., Laj, C., de Seze, L. & Verosub, K.L., 1983 *15 Myr periodicity in the frequency of geomagnetic reversals since 100 Myr.* Nature **304**, 328.

McCrea, W.H., 1975 *Ice ages and the Galaxy.* Nature **225**, 607.

McCrea, W.H., 1981 *Long time-scale fluctuations in the evolution of the Earth.* Proc. Roy. Soc. London A, **375**, 1.

McLaren, D.J., 1970 *Time, life and boundaries.* (presidential address). J. Paleont **44**, 801.

Muller, R.A. & Morris, D.E., 1986 *Geomagnetic reversals from impacts on the Earth.* Geophys. Res. Lett. **13**, 1177

Napier, W.M., 1987 *The Origin and Evolution of the Oort cloud.* In INTERPLANETARY MATTER (eds. Z. Ceplecha & P. Pecina, Proc. Tenth European Regional Meeting in Astronomy, Volume, 2 Prague), p 13.

Napier, W.M. & Clube, S.V.M., 1979 *A theory of terrestrial catastrophism.* Nature **282**, 455.

Napier, W.M. & Staniucha, M., 1982 *Interstellar Planetesimals. I. Dissipation of a primordial cloud of comets by tidal encounters with massive nebulae.* Mon. Not. R. astr. Soc. **198**, 723.

Negi, J.G. & Tiwari, R.K., 1983 *Matching long-term periodicities of geomagnetic reversals and galactic motions of the solar system.* Geophys. Res. Lett. **10**, 713.

O'Keefe, J.D. & Ahrens, T.J. 1982 *Impact mechanics of the Cretaceous-Tertiary bolide.* Nature **298**, 123.

Oort, J.H., 1950 *The structure of the cloud of comets surrounding the solar system and a hypothesis concerning its origin.* Bull. Astr. Inst. Neth. **11**, 91.

Öpik, E.J., 1963 *Stray bodies in the solar system. Part 1. Survival of cometary nuclei and the asteroids.* Adv. Astron. Astrophys. **2**, 219.

Öpik, E.J., 1958 *On the catastrophic effects of collisions with celestial bodies.* Irish Astron. J. **5**, 34.

Pal, P.C. 1988 Personal communication.

Pandey, O.M. & Negi, J.G., 1987 *Global volcanism, biological mass extinctions and the galactic vertical motion of the solar system.* Geophys. J. R. astr. Soc. **89**, 857.

Pavlovskaya, E.D. & Suchkov, A.A., 1980. *Galactic spiral structure parameters : error estimates by numerical experiments.* Sov. A.J. **24**, 164.

Rampino, M.R. & Stothers, R.B., 1984 *Terrestrial mass extinctions, cometary impacts and the Sun's motion perpendicular to the galactic plane.* Nature **303**, 709.

Raup, D.M., 1986 *Biological extinction in Earth history.* Science **231**, 1528.

Raup, D.M. & Sepkoski, J.J., 1984 *Periodicity of extinctions in the geologic past.* Proc. Nat. Acad. Sci. U.S.A. **81**, 801.

Raup, D.M. & Sepkoski, J.J., 1986 *Periodic extinctions of families and genera.* Science **231**, 833.

Rayleigh, Lord, 1880. Phil. Mag. **10**, 73.

Russell, D.A., 1979 *The enigma of the extinction of the dinosaurs.* Ann. Rev. Earth Planet. Sci. **7**, 163.

Sepkoski, J.J., 1989 *Periodicity in extinction and the problem of catastrophism in the history of life.* J. Geol. Soc. London **146**, 7.

Sepkoski, J.J. & Raup, D.M., 1986 *Periodicity in marine mass extinctions.* In DYNAMICS OF EXTINCTION (ed. D. Elliott), p.3. J. Wiley and Sons, New York.

Seyfert, C.K. & Sirkin, L.A., 1979 EARTH HISTORY AND PLATE TECTONICS Harper and Row, New York.

Shapley, H., 1921 *Note on a possible factor in changes of geological climate.* J. Geol. **29**, 502.

Shoemaker, E.M. & Wolfe, R.F., 1986 *Mass extinctions, crater ages and comet showers.* In THE GALAXY AND THE SOLAR SYSTEM (eds. R. Smoluchowski, J.N. Bahcall and M.S. Matthews), p.338. University of Arizona Press, Tucson, Arizona.

Smit, J. & Hertogen, J., 1980 *An extraterrestrial event at the Cretaceous-Tertiary boundary.* Nature **285**, 198.

Stagg, C.R. & Bailey, M.E.B., 1989 *Stochastic capture of short-period comets.* Mon. Not. R. astr. Soc., in press.

Steiner, A. & Grillmair, E., 1973 *Possible galactic causes for periodic and episodic glaciations.* Geol. Soc. Amer. Bull. **84**, 1003.

Stothers, R.B., 1986 *Periodicity of the Earth's magnetic reversals.* Nature **322**, 444.

Stothers, R.B., 1988 *Structure of Oort's comet cloud inferred from terrestrial impact craters.* The Observatory **108**, 1.

Stromgren, B., 1967 *In Radio Astronomy and the Galactic System.* IAU SYMP. NO. 31 (ed. H. van Woerden), p.303, Academic Press, New York.

Talbot, R.J., 1980 *Rate of star formation dynamical parameters, and interstellar gas density in our Galaxy and M83.* Astrophys. J. **235**, 821.

Thaddeus, P. & Chanan, G.A., 1985 *Cometary impacts, molecular clouds, and the motion of the Sun perpendicular to the galactic plane.* Nature **314**, 73.

Trefil, J.S. & Raup, D.M., 1987 *Numerical simulations and the problem of periodicity in the cratering record.* Lett. **82**, 159.

Tremaine, S., 1986 *Is there evidence for a solar companion star?* In THE GALAXY AND THE SOLAR SYSTEM (eds. R. Smoluchowski, J.N. Bahcall and M.S. Matthews), p.409. University of Arizona Press, Tucson, Arizona.

Urasin, L.A., 1987 *Spiral model of the Galaxy from observations of interstellar extinction.* Sov. Astron. Lett. **13**, 356.

Urey, H.C., 1973 *Cometary collisions and geological periods.* Nature **242**, 32.

Wdowczyk, J. & Wolfendale, A.W., 1977 *Cosmic rays and ancient catastrophes.* Nature **268**, 510.

Whitmire, D.P. & Jackson, A.A., 1984 *Are periodic mass extinctions driven by a distant solar companion?* Nature **308**, 713.

Woolley, R.v.d.R. 1965 *Motions of the nearby stars.* in GALACTIC STRUCTURE (eds. Blaauw A. & Schmidt M.), p 85 University of Chicago Press.

Yabushita, S., 1979 *A statistical study of the evolution of the orbits of the long-period comets.* Mon. Not. R. astr. Soc. **187**, 445.

Yabushita, S., 1989 *On the discrepancy between supply and loss of observable long-period comets.* Mon. Not. R. astr. Soc., in press.

Zhao, M. & Bada, J.L. 1989 *Extraterrestrial amino acids in Cretaceous-Tertiary boundary sediments at Stevns Klint, Denmark,* Nature **339**, 463.

METEOROID STREAMS AND THE ZODIACAL DUST CLOUD

Duncan Olsson-Steel

Department of Physics and Mathematical Physics,
University of Adelaide, G.P.O. Box 498
Adelaide, SA 5001, Australia

Summary. The origin and evolution of meteoroid streams, the way in which these produce the sporadic meteoroids, and the production of the zodiacal dust from the comminution of meteoroids, are all discussed with reference to how the present situation may reflect the events around the epoch of a large increase in the flux of comets passing through the inner Solar System, such as has been suggested as causing mass extinctions of terrestrial lifeforms in the past. Appropriate time-scales governing the processes affecting the small bodies under consideration here (including the production of meteoroids by comets and asteroids, the disruption of streams, physical decay of meteoroids to form the zodiacal dust, and the loss of dust from the complex through inspiralling towards the Sun) are of order 10^3–10^5 yr, so that present-day studies of meteoroids and dust can give us important information about the microstructure of astronomically-induced terrestrial catastrophes. It is contended that, in line with suggestions by other authors, the present complex of inter-planetary material reflects a phase of gradual decay after the break-up of a Giant Comet some time in the past $\sim 10^4$ yr. It is also suggested, from the evidence of the available meteoroid orbit data, that there is a substantial population

of planet-crossing asteroids and short-period comets in high-inclination/retrograde orbits which wait discovery: such bodies would be of enormous importance with regard to the rate of impacts of large objects upon the Earth.

Introduction

In this chapter some features of the origin and evolution of the smaller bodies in the inner Solar System (sizes below about 10cm–1m) will be discussed with particular reference to how these particles may give us clues to the variation in the environment of this region on astronomically-brief time-scales (10^3–10^5 yr). Such studies are thus of importance in understanding how the climatic variation on the Earth over the same period might be linked to the influx of such bodies to our planet, and hence what might have occurred on a similar time-scale around the epochs in which the cometary flux in the inner Solar System was greatly increased (*i.e.* at the times of cometary waves which have been postulated as causing mass extinctions on the Earth, as discussed in other contributions to this volume). Although the present paper concentrates upon meteoroids and zodiacal dust, these particles are derived directly from larger bodies (asteroids and comets), so that some discussion of the linkage between the various classes of objects is necessary; however, for more detailed discussions of the origin and dynamics of comets and asteroids, and the ways in which these large bodies may affect life on the Earth directly through massive impacts, the reader is again referred to other contributions.

There is now a realization that the overall complex of planet-crossing bodies is derived at least in part from the influx of long-period (LP: period $P > 200$ yr) comets which are captured into short-period (SP: $P < 200$ yr) orbits by planetary perturbations (for reviews see Weissman 1985 and Bailey *et al.* 1986). Until recently it had been believed that SP comets were the major source of interplanetary dust and small meteoroids (Olsson-Steel 1986) but it has now been recognized that LP comets can make a major direct contribution to the population of smaller particles (Fulle 1987, 1988), and also that there are meteoroid streams associated with several Apollo asteroids which may be either collisional debris or remnant tails from when the asteroids were active cometary nuclei (Olsson-Steel 1988a). The physical decay of comets, with concomitant production of meteoroids and dust, has been the subject of much study (*e.g.* Kresák 1981, Fernández 1985), as has the orbital evolution of SP comets and their dynamical relationship to planet-crossing asteroids (Carusi *et al.* 1985, Belyaev *et al.* 1986, Hahn & Rickman 1985). The physical relationship between Earth-approaching asteroids and comets has also been the subject of many investigations — see Degewij & Tedesco (1982) and Hartmann *et al.* (1987) for reviews. In turn the meteoroids released maintain the zodiacal dust cloud through catastrophic collisions (Whipple 1967, Leinert *et al.* 1983, Grün *et al.* 1985, Steel & Elford 1986, Olsson-Steel 1986), whilst it has recently been discovered that there are various other mechanisms and sources which can help replenish this cloud (*e.g.* Sykes & Greenberg 1986, Gustafson *et al.* 1987, Sykes 1988). Thus in order to understand the terrestrial effects of a wave of LP comets from the Oort Cloud, as has been postulated as being the direct cause of mass extinctions of fauna (*e.g.* Hut *et al.* 1987, Stothers 1988, see also Bailey *et al.* 1987), it is essential that we have a good understanding of the ways in which a LP comet, being deviated from a large orbit when some perturbation inserts it into a path with small perihelion distance, evolves to become a SP comet and then (possibly) a dark asteroid, producing meteoroids and dust in the process (Hughes 1985, Rickman 1985).

To set the mood for this paper, it would be useful to mention some of the questions which would be pertinent in this respect:

- What is the origin of meteoroid streams? Are they all derived directly from comets, so that the asteroid-related streams discussed in the next section are evidence for these parent bodies being extinct or moribund comets, or can streams be formed as the result of impacts upon the asteroids by boulder-sized objects?

- What is the origin of the sporadic meteoroids? Are these derived from the dispersal of streams, and if so what are the appropriate time-scales?

- What limits the lifetimes of meteoroids? Do inter-particle collisions lead to significant losses of stream meteoroids or the production of sporadics? Is the bulk of the zodiacal dust cloud (particle sizes $\sim$10–100μm) produced by such catastrophic collisions?

- What is the time-scale for the loss of zodiacal dust particles (zdp's) under the Poynting-Robertson effect? Does this allow us to further constrain the ways in which the interplanetary flux of comets/asteroids/meteoroids has changed over the past 10^3–10^5 yr?
- Is the population of meteoroids/zdp's in a steady-state? Does the possible lack of a steady-state add weight to the recent Giant Comet hypothesis of Clube & Napier (1982, 1984, 1986)?
- Does the existence of certain individual objects which appear to be genetically-related also favour the possibility of a recent Giant Comet?
- If some significant fraction of planet-crossing asteroids are in fact remnant cometary cores, then might we not expect there to be a population of such asteroids in high-inclination (possibly even retrograde) orbits? Does the existence of many retrograde meteoroids argue for the existence of many parent bodies (Apollos/undiscovered comets?) in such orbits? If these do exist but remain as yet undiscovered, then what might their effect be upon the terrestrial large-body impact rate?
- During a cometary wave the flux of meteoroids entering the Earth's atmosphere would be expected to be much higher than the long-term average. How might this affect the climate, in view of our knowledge of the importance of meteoric species in contemporary high-altitude aeronomy? Could the meteoric influx at such times cause a more significant perturbation to the terrestrial environment than would the few large-body impacts which have been suggested as causing widespread faunal extinctions? (*cf.* Clube 1987, where this possibility is considered in relation to the evolution of individual Giant Comets).

The origin of meteoroid streams

The fact that many meteoroid streams, observed on the Earth as meteor showers, are derived from particular comets has been established for over a century (*e.g.* Porter 1952; Hughes 1978). Specific cases include Comet P/Swift-Tuttle and the Perseid shower of August, and Comet P/Halley and the Eta Aquarids of May and the Orionids of mid-October: the various comet-shower associations have been discussed by Drummond (1981). The streams consist of the larger solid particles ejected by the comet as it approaches perihelion, with the meteoroids being swept away from the nucleus by the expanding gaseous coma produced by the evaporation of ice due to heating by solar radiation (Hughes 1985).

It is well-known that meteor showers are more noticeable amongst brighter meteors in the Earth's atmosphere (*i.e.* larger mass meteoroids), and this is partly due to the fact that such bodies have smaller ejection velocities from the comet and thus are not so widely dispersed in their orbital elements as are the low-mass particles: thus there is little evidence of distinct showers in data collected using, say, a meteor radar with a low limiting magnitude, with almost all meteors occurring semi-randomly (*i.e.* sporadic meteors); however, a case has been made for very small particles being present in some streams (Singer & Stanley 1980). After release from the parent comet, the meteoroids have slightly different heliocentric velocities and soon spread around the cometary orbit so as to form a complete loop: this occurs typically on a time-scale of a few tens of orbital periods (*e.g.* Williams 1985). Occasionally a young comet which has not yet had a chance to form a complete stream loop may cross the ecliptic near 1 AU with the Earth reaching that point at much the same time: under such circumstances a meteor storm may occur, such as the periodic Leonids or October Draconids/ Giacobinids (Yeomans 1981, Williams *et al.* 1986).

After release from the comet the meteoroids do not continue on the same orbits in perpetuity but are gradually dispersed by a number of agents. Having a variety of semi-major axes the meteoroids suffer differential planetary perturbations, which can lead to some spreading although by no means enough to explain the duration of meteor showers, which indicate stream widths of order at least 0.1–0.2 AU. Thus although planetary perturbations appear to cause the rotation of the angular elements of streams (*e.g.* they can explain why the Geminid shower was not observed prior to the nineteenth century:

Hunt *et al.* 1985), other influences are required to explain the wide dispersal in elements observed in some streams. Particular streams of interest are the Geminids and the Perseids: the former has its aphelion well within Jupiter's orbit and thus suffers insignificant dispersion by planetary perturbations, whilst the latter, although a crosser of the outer planets, actually has an orientation which prohibits close approaches to those bodies, so that again some other dissipative force seems necessary. Olsson-Steel (1987a) has argued that the dispersion comes about from the combined effect of the radiative forces acting upon individual meteoroids, to wit radiation presssure, Poynting-Robertson effect and the Yarkovsky-Radzievskii effect, with the last (which depends upon the particle's spin rate and direction, along with other physical properties) being especially significant. This suggestion has been criticized (Babadzhanov & Obrubov 1987), but the author takes this opportunity to reiterate his conviction that the dispersal of meteoroid streams proceeds mainly through differing orbital energies being produced predominantly by the radiative effects (rather than differing ejection speeds and directions from the parent object), with differential planetary perturbations then causing the variations in the other orbital elements, namely argument of perihelion, longitude of node, and inclination.

Over the past few years physical studies of the smaller bodies in the inner Solar System, as mentioned in the Introduction, have led to a blurring of the distinction between comets and asteroids, which was historically based upon their telescopic appearance (hence the derivation of these particular words). Öpik (1963) suggested that the Earth-crossing (Apollo) asteroids are in fact remnant cometary cores, this idea being based upon the realization that these objects could not have existed in their present orbits since the Solar System was formed 4.6 billion years ago since their lifetimes against collisions with the terrestrial planets are only of order 10 million years. More recent work has shown that an origin in the asteroid belt might be possible (Wisdom 1985; Wetherill 1985), although this does not yet seem secure. The discovery in 1983 of 3200 Phaethon, which is apparently the parent of the Geminid meteoroid stream but nevertheless appears asteroidal in nature, strongly pointed towards the evolution of comets into asteroids, and recent work reported by Olsson-Steel (1988a) and also discussed here has identified meteoroid streams associated with several other Apollo asteroids.

Although the comet/meteor-shower link has been known for many years, despite several suggestions of asteroid/shower associations (*e.g.* Hoffmeister 1948, Sekanina 1973, 1976; Babadzhanov & Obrubov 1983) none has been generally accepted until Phaethon was found and recognized by Whipple (1983) to be in an orbit virtually identical to that of the Geminids. One of the problems inherent in recognizing associations between meteoroid streams in small orbits (as compared to the larger comet-related streams in such as the Halleyids, Perseids or Leonids) and their parent objects is that the majority of meteoroids have aphelia within about 5 AU (*e.g.* see the plots for the Adelaide data presented by Olsson-Steel (1988a) or for all available surveys by Olsson-Steel & Lindblad (1989)), so that there is typically a strong background of sporadics with similar orbital elements, and it is difficult to prove the existence of distinct streams. Objects on prograde orbits which approach Jupiter are typically ejected from the Solar System or diverted into radically different orbits on time-scales of order $\sim 10^4$ yr or less (Olsson-Steel 1986, 1988b; see later) so that streams are unrecognizable after such a length of time, with these deflected meteoroids apparently maintaining the sporadic background. However streams in smaller orbits will live for a longer period, although gradual dispersal as described above will cause substantial broadening, with the lifetime being limited by the loss of individual meteoroids in collisions with zdp's. This means that we might expect to find Apollo-related streams which are of low flux (since many of the meteoroids may have been lost) but are still detectable in a data set of suitable orbits, given an appropriate search technique.

The method used here was more fully described by Olsson-Steel (1988a). Briefly, the conventional way of searching for meteoroid streams in large sets of orbit data has been to use some 'seed orbit' (*e.g.* a suspected stream or parent body orbit) and then compare that orbit with the meteor data using the so-called D–Criterion of Southworth & Hawkins (1963), or the revised criterion (D′) of Drummond (1981). However, this can lead to problems since it is not known how many meteoroids may appear to be members of a particular stream but in fact are just sporadics which happen to have similar orbital parameters to the stream under investigation. The problem is quite complicated since the search is being done in a multi-dimensional phase-space (*i.e.* five orbital elements a, e, i, ω, Ω, with the time of perihelion passage also being involved if the stream is not evenly dispersed around the orbit), and the density of orbits within each dimension would be required in order to known whether a stream is really present (*i.e.* whether there is a number of meteors with similar elements which is significantly higher than the random, sporadic distribution). Even then, what about the case of an old, highly-dispersed stream? How does one define when a stream ceases to exist as such and its members are part of the sporadic population?

A rather simpler approach has therefore been used, which can be summarized as follows. A meteor shower consists of a number of meteors with a similar radiant and velocity (which means similar orbital elements) which appears each year at the same time (*i.e.* at the same solar longitude). Thus, if one uses a seed orbit in a search with a certain (a, e, i, ω) but with the longitude of the ascending mode Ω taking all values from 0° to 360° then a concentration of the number of similar meteor orbits near to a particular Ω shows the existence of a stream. This method assumed that there is a fairly even coverage of detected meteors over the whole year, and also that the Earth's orbit is sufficiently close to being circular that the orbits being searched for are not excluded (*e.g.* if one were searching for a stream with perihelion distance $q = 1.015$ AU then this could only be detected for a short time each year, and so would automatically lead to a concentration near that time). Another point of caution is that the radiant for a particular set of elements (a, e, i, ω) has a declination which varies with Ω due to the obliquity of the ecliptic: thus for a survey conducted from Adelaide at 35° south latitude, and capable of detecting meteors to 40° north declination, meteors with radiants at $\delta = 35°$ would be detectable but the same orbital elements might lead to $\delta = 45°$ at another time of year, and the meteors would not be accessible from that site.

Despite these minor drawbacks the method used here proves to be an excellent and easily applied technique which is capable of unequivocably demonstrating the presence of asteroid-related meteoroid streams, and has been used with the Adelaide radar meteor orbit surveys by Olsson-Steel (1988a) to show that at least some, and perhaps many, of the Apollo-type asteroids have associated streams: in particular 1566 Icarus, 2101 Adonis, 2201 Oljato, 2212

Hephaistos, 3200 Phaethon, 1937 UB (Hermes), 5025 P–L, 1982 TA and 1984 KB. This work is now being extended to include the meteor data available from various other surveys conducted in the U.S.A., U.S.S.R., Ethiopia and Canada; generally the deductions based on the Australian data are confirmed. As examples we show here the results for Apollo-type asteroids 1984 KB and 1566 Icarus. Figure 1 shows the plots for 1984 KB of the number of correlated meteors as a function of nodal longitude, Ω. The Adelaide data are those described by Nilsson (1964), Gartrell & Elford (1975) and Olsson-Steel (1988a); the Obninsk data were discussed by Lebedinets *et al.* (1981, 1982); and the Harvard data by Sekanina (1973, 1976). Each plot shows a peak which occurs close to the actual Ω of 1984 KB, indicating that not only does a stream exist but also it is apparently derived from this parent object. At more than 30–40° from the Ω of 1984 KB, the counts will be due to sporadic meteors; although there are appreciable numbers of these it is due to the fact that the orbit of this particular asteroid is quite similar to the bulk of the sporadic meteor population (see the orbital element distribution plots presented by Olsson-Steel (1988a) and Olsson-Steel & Lindblad (1989)). It is noticeable that the plots for the two Harvard surveys show rather broader peaks than for the Adelaide and Obninsk data: this is apparently due to the fact that the Harvard data relate to rather fainter meteors (smaller meteoroids), for which the perturbational forces in space (radiative effects, larger ejection velocities from the parent) lead to such particles becoming dispersed from their original orbits much more quickly than the larger meteoroids detected from Adelaide and Obninsk, and hence form a broader stream.

Note that 1984 KB appears to be a member of the Taurid complex of objects, which includes various comets, asteroids, fireballs and meteor showers (Clube & Napier 1984, 1986; Olsson-Steel 1987b, 1988c); this complex is discussed in a later section. Exactly the same sort of plots are shown in Figure 2 for the case of 1566 Icarus. This asteroid had previously been suggested as being the parent of a meteor shower (see Sekanina 1973, 1976) but this link has not been generally accepted. The evidence for a stream derived from Icarus based upon all available meteor orbit data will be dealt with in another paper, but a few brief notes are included here.

The strongest of the daytime (radar-observable) meteor showers

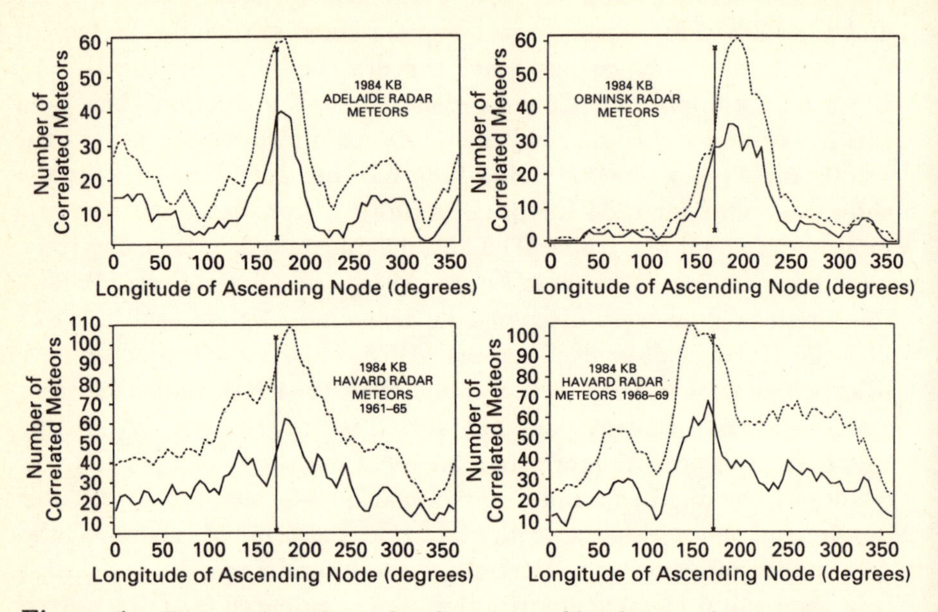

Figure 1. **The number of correlated meteor orbits from various surveys as a function of the assumed Ω for 1984 KB; the vertical line at 170° is the actual Ω of this asteroid. The solid line is for D < 0.20 (Southworth & Hawkins, 1963) and the dashed line for D′ < 0.125 (Drummond 1981).:**

is called the Daytime Arietids; these recur in late May/early June each year. In Table I we compare the mean orbital elements of the meteoroids in this shower with the elements of Icarus. Clearly there is a marked similarity between the two sets of elements, and it is found that most of the orbits contributing to the peaks in Figure 2 are in fact members of the Daytime Arietid stream. If in place of the orbit of Icarus the stream orbit in Table 1 is used as the seed in the search for correlated meteors, then it is found that similar plots to those in Figure 2 are obtained, except that the peaks are even stronger and also these peaks occur right on the input nodal longitude ($\Omega = 77°.0$) rather than displaced by a few degrees as when $\Omega = 87°.52$ is used. Overall it appears that what may be occurring here is that, at the present time, the Earth is passing through the edge of the stream, such that the meteoroids detected have orbits which are distinct from the parent, since they have suffered various perturbations which have pushed them to the stream periphery. This

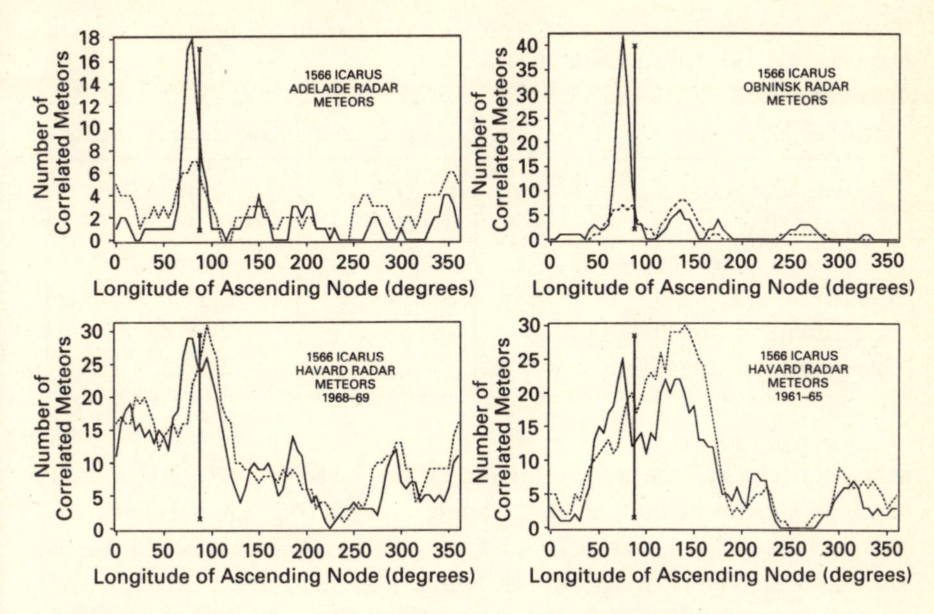

Figure 2. **As Figure 1 except for 1566 Icarus. Note that for the Harvard Surveys, which detected much fainter meteors (smaller meteoroids), the stream is much broader (*i.e.* the shower continues for a much longer time) since the dispersion in space is much swifter for smaller bodies.:**

then opens up the question of how the shower characteristics may vary with time: as ω and Ω rotate under planetary perturbations, the activity and times of occurrence of the shower may be expected to change. This being a daytime shower, its existence has only been known for forty years, unlike the Geminids which have been observed since their appearance in 1862. Certainly, radar monitoring of the activity of the Daytime Arietids over the next few decades is warranted, as are orbital evolution studies like that carried out by Hunt *et al.* (1985) for the Geminids.

One further point is that the perihelion distances of the stream and the asteroid are substantially different. Although individual radar meteor orbit determinations are notoriously inaccurate, the mean shower orbit is well-established and definitely has q = 0.08–0.10 AU. If the stream and the asteroid are genetically related then this requires an explanation. Possibly the simplest (but merely a get-out) would be to postulate that the asteroid, whilst an active comet,

Table 1. Orbital elements of Apollo-type asteroid 1566 Icarus and the Daytime Arietids (from Cook 1973):

Orbital Elements	1566 Icarus	Daytime Arietids
Semi-major axis, a	1.0079 AU	1.50 AU
Eccentricity, e	0.8268	0.94
Perihelion distance, q	0.1867 AU	0.09 AU
Inclination, i	22°.90	21°.0
Argument of perihelion, ω	31°.16	29°
Longitude of ascending node, Ω	87°.52	77°.0

orginated the stream at some time in the last 10^3–10^4 yr (probably towards the lower end of that range, considering the strength and compactness of the shower), but has since by chance suffered a close approach to Venus or the Earth which has diverted it into an orbit with larger q itself, but left most of the meteoroids on the original small-q orbits. Certainly orbital evolution integrations would be of interest in this respect; however, if such an encounter had occurred then the other elements would be expected to alter quite quickly so that any similarity disappears. An alternative possibility is that rather than being a remnant cometary tail, this stream in fact came about as the result of one or more impacts upon the asteroid whilst it was near aphelion. This would be the point at which most impacts by boulder-sized objects would be expected to occur, since aphelion occurs at the inner edge of the asteroid belt. Such collisions would be expected to result in the release of meteoroids into orbits which are basically the same but have differing perihelion distances (*i.e.* all particles return to to the same point on the next orbit — that is, aphelion — but their dispersion in velocity must cause their perihelion distances to vary.)

The origin of sporadic meteors

The origin of the sporadic meteoroid component was a long-standing problem of meteor astronomy which engaged the occasional attention of researchers for some years. Although the limiting lifetime of meteoroids in the 100μm–10cm size range is known to be due to collisions with smaller zodiacal dust particles (sizes 10–100μm), as was shown by Dohnanyi (1978, and earlier papers cited therein), such collisions do not lead to the orbital elements of the fragments becoming widely dispersed so that this is not the mechanism by which the sporadic orbits originate. Olsson-Steel (1986) has compared the time-scales for different influences upon the orbits of meteoroids (loss through collisions with zdp's, collisions with the planets or ejection from the Solar System, orbital collapse under the Poynting-Robertson effect, randomization of the orbits due to close encounters with the giant planets, in particular Jupiter) and shown that the time-scale for such randomization (i.e. becoming a sporadic rather than a stream meteoroid) is much shorter than the loss time-scales for prograde orbits with aphelia near Jupiter; for such orbits the encounters with the giant planet are quite frequent and are also efficient in bringing about orbital change since the relative velocity is low. Typical limiting lifetimes for 1mm meteoroids vary from a few times 10^4 to 10^6 yr, being near 10^5 yr for prograde orbits with Q $\approx$ 5 AU, whereas sporadic production occurs on a time-scale of 10^3–10^4 yr. Such orbits are also similar to the bulk of the SP comets (Marsden 1986) since Jupiter is responsible for capturing these bodies from LP orbits, and in addition the majority of known meteoroid streams have orbits of this type (Cook 1973). Thus, it appears that sporadic meteors are produced from streams on a time-scale of order 10^3–10^4 yr from comets with Q $\approx$ 5 AU and with low inclinations.

The point of the last paragraph is that this implies that a study of the balance between sporadic and stream meteors in principle allows an investigation of several features of the recent history of meteoroid production and hence the influx of SP comets, which are themselves derived from original LP orbits. If there has been a major increase in the number of meteoroids (initially in streams) around 10^5 yr ago then we would now expect to see about e^{-1} of these remaining, the rest have been ground up into zdp's, and almost all of the remnant meteoroids would now be sporadics since they become randomized on a time-scale at least an order of magnitude shorter. In fact the fraction of meteoroids in streams is usually estimated as being somewhere between 10 and 50% of the total flux (see Hughes 1978, Olsson-Steel

1986) depending on the criteria used; and if the technique already described were applied to all measured meteor orbits when a larger fraction of the possible parent objects have been discovered, then it would be expected that in fact over 50% of meteroids may turn out to be stream members. This, then, implies that the current population of meteoroids have largely been released after an enhancement in their production rate sometime in the past $\sim 10^4$ yr.

Production of zodiacal dust particles

If, as is suggested by the present results concerning the origin of meteoroid streams, the Apollo asteroids are a major source of meteoroids, then this helps to solve one of the outstanding problems in the ecology of interplanetary objects: the apparent shortage of parents sufficient to explain the population of these bodies (*e.g.* see Whipple 1967, Olsson-Steel 1986, Fulle 1987, Clube 1987). The objects observed in the Earth's atmosphere as meteors (*i.e.* sizes from 100μm to several centimeters) are known to power the zodiacal dust cloud (particle sizes mostly in the range 10–100μm) through catastrophic collisions (see Grün *et al.* 1985, and other papers cited in the Introduction) but, as Grün *et al.* have shown, there seems to be a surfeit of meteoroids, by about an order of magnitude. The time-scale for the loss of zdp's due to inspiralling towards the Sun under the influence of Poynting-Robertson (P–R) effect with eventual loss due to mutual collisions or evaporation is required. The estimation of this time is quite straightforward (*e.g.* see the review by Burns *et al.* 1979), and the calculations of Olsson-Steel (1986) have shown that for millimetre-size meteoroids on typical stream orbits the appropriate time-scale is generally near 10^6 yr although it can be as low as $5 \times 10^4 - 1 \times 10^5$ yr for orbits with small perihelion distances, such as the Geminids, Daytime Arietids, or Delta Aquarids. Since the P–R lifetime varies as the relative size this means that for 10μm particles the appropriate time-scale is $\sim 10^4$/yr but may be 10^3/yr or less for particles with small q. Thus although some dust particles (the smaller zdp's with small perihelion distances) are lost on P–R time-scales which are shorter than their production rate from the comminution of meteroids, in general the dust remains for longer periods. This argues for a non-steady-state, with the present epoch reflecting a phase of decay of the number of meteoroids (or a build-up in the amount of dust) after a recent large enhancement, in line with the earlier discussion of the sporadic/stream meteoroid balance.

This section and that previous have indicated that in the present epoch the population of small solid bodies in the inner Solar System may not be in a steady-state, but the phenomena observed now may represent a gradual decay in the total mass of material after a recent enhancement (time-scale $\sim 10^4$ yr). Additional evidence for this is the difficulty in explaining the number and distribution of SP comets, if these are derived directly from the LP comet flux observed now (Everhart 1973, Weissman 1985, Duncan *et al.* 1988, Stagg & Bailey 1989).

The Giant Comet hypothesis

The key to this situation may be the Taurid complex of interplanetary objects, which includes Comet P/Encke (previously suggested as the major source of the zodiacal dust cloud: see Whipple (1967), Clube & Napier (1984), Štohl (1986) & Gustafson *et al.* (1987)), four of the Apollo asteroids with associated meteoroid streams, and also possibly Comet 1967 II Rudnicki (Napier 1983, Olsson-Steel 1987b). The existence of these apparently-related objects argues for the breakup of a single progenitor, suggested as being a Giant Comet by Clube & Napier (1982, 1984, 1986). Additional evidence for this has recently been outlined by Olsson-Steel (1988c), who points out that if the Brno fireball observed in 1978 was in fact another member of the complex then the size of the Giant Comet may be estimated at ~ 40 km. Ziołkowski (1988) has re-analysed the astrometric observations of Comet 1967 II Rudnicki and shown that after allowance for non-gravitational effects and reference frame transformation the original orbit appears to have been elliptical, with a period of order a few times 10^4 yr, again adding weight to the idea that a Giant Comet may have entered the inner Solar System around that time. Various other objects which may be part of the complex have been listed by Clube & Napier (1986), including the Tunguska object (see also Kresák 1978).

A high-inclination Apollo asteroid/SP comet population

In keeping with the general philosophy of this paper in which various speculations and pertinent questions have been aired, in this final section we make some suggestions concerning the possibility that there exists an as yet undiscovered population of high-inclination planet-crossing asteroids and short-period comets. This radical suggestion is prompted by a consideration of the orbital distribution of meteoroids, which require parent objects with similar orbital characteristics.

Figure 3 shows the inclination distributions for the 613 LP comets and 135 SP comets listed by Marsden (1986), the 85 Aten-Apollo-Amor asteroids discovered through to 1986 October (as used by Olsson-Steel (1988a) with their orbital elements being listed by Olsson-Steel (1987c)), and the 3759 meteor orbits measured at Adelaide during the 1960's (Nilsson 1964, Gartrell & Elford 1975). Before making any further comments it is important to emphasize that all four distributions are biassed samples, due to the differing discovery

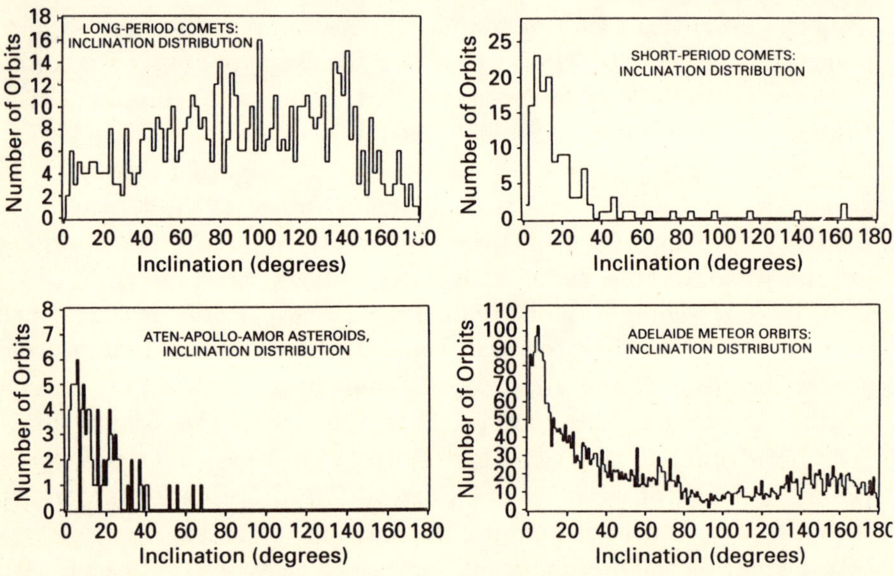

Figure 3. **The distribution of the inclinations of all known long-period comets, short-period comets, Aten-Apollo-Amor asteroids and the radar meteor orbits measured in the Adelaide surveys.:**

probabilities for the large bodies, and for the meteors due to various selection effects. In fact the sampling bias is especially severe for the meteors, since in order for a meteoroid to be detected it must (i) collide with the Earth, with different Earth-crossing orbits having different collision probabilities; (ii) have a radiant which is within the radar beam pattern of the detection system, with each system having a pattern which is heavily declination- (and hence orbit-) dependent; (iii) give sufficient ionization to be detected by the radar, this being a strong function of the meteor velocity and hence the original orbit; and (iv) be at the altitude which is amenable to meteor detection with a VHF radar: Olsson-Steel & Elford (1987) have discussed this in detail, but it is worthwhile pointing out here that meteors ablating above about 105 km are unlikely to be detected with a VHF radar, and such higher-altitude meteors are likely to have originally been of lower density (i.e. cometary rather than asteroidal/rocky in nature). Thus, it should not be imagined that the plots in Figure 3 are by any means indicative of the real interplanetary distribution: in fact the point of the present section is to make a case for the real (unbiassed)

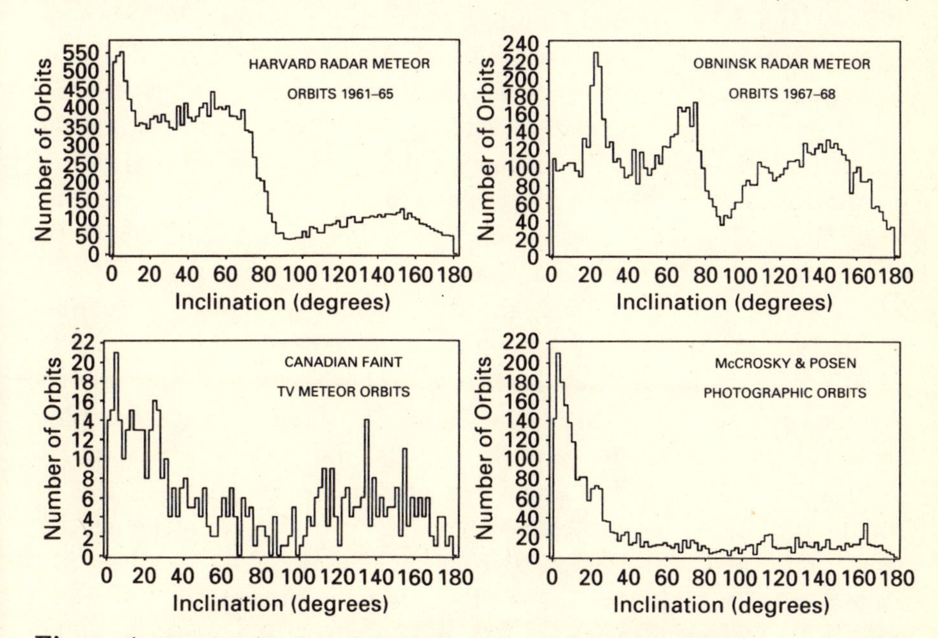

Figure 4. The distribution of the inclinations of the meteoroid orbits measured in the Harvard 1961–65 radar survey, the Obninsk radar survey, the Canadian faint TV meteor survey and the Harvard photographic survey.:

SP comet and AAA asteroid distributions containing a larger number of higher-inclination objects.

In order to emphazise the fact that meteor orbit surveys are highly selective, but nevertheless a substantial proportion of retrograde orbits are measured in each, we also show in Figure 4 the inclination distributions from the Harvard Radar Meteor Project of 1961–65 (Sekanina 1973: 19327 orbits), the Obninsk radar survey (Lebedinets *et al.* 1981, 1982: 9358 orbits), the Canadian faint TV meteor surveys (Hawkes *et al.* 1984, Jones & Sarma 1985: 531 meteors), and the Harvard photographic orbits (McCrosky & Posen 1961: 2414 orbits). Inclination distributions for other surveys are presented by Olsson-Steel & Lindblad (1989). Since the plots in Figure 4 are representative of differing limiting meteor magnitudes (and hence various masses), one would not necessarily expect the distributions to be identical, but we note that the fraction of detected meteoroids which are in retrograde (inclination $i > 90°$) orbits varies from 10–20% for the Harvard radar and McCrosky & Posen photographic meteors, to 30–40% for the Adelaide and Obninsk radar and Canadian TV meteors. These are realised to be extremely high fractions when one takes into account the fact that the collisional lifetime of meteoroids in space varies basically as the relative velocity between themselves and the target material (the zodiacal dust, which is in low-eccentricity, low-inclination orbits): thus, since, this velocity is at least two or three times higher for the retrograde meteors, their population should fall off at a rate e^2 - e^3 times faster than the prograde meteors (see Steel & Elford 1986). Therefore the high proportion of retrograde meteoroids argues for both a high production rate, and also a recent enhancement.

We now compare these meteor distributions with those for the macroscopic (parent) objects in Figure 3. First, the meteor plots (with the possible exception of the Obninsk data) show large numbers of orbits with $i < 20°$. These are similar to the peaks in the SP comet and AAA asteroid plots, backing up the general belief that most meteoroids are liberated from such bodies (see Hughes 1978; Fulle 1987, Štohl 1986; Olsson-Steel 1988a). However there are no known AAA asteroids with $i > 70°$, and only five SP comets (including P/Halley) with retrograde orbits, these all having aphelia well beyond Jupiter (see the orbits listed in Marsden (1986); also Olsson-Steel (1988d) for a discussion of the origin of these comets). Even if there were twice as many SP comets of this type this would still not explain the number of retrograde meteoroids, since (i) there would be a problem with the supply rate; and (ii) there is a semi-continuous distribution of retrograde meteoroids, whereas several distinct peaks would be expected if there were a small number of parent bodies (*cf.* the peak near $i = 164^o$ in the McCrosky & Posen orbits, this being due to the Halleyids, with the Perseids at 114°–116°).

One therefore turns to the LP comets as the source of the high inclination meteoroids. The LP comet inclination distribution shows many retrograde orbits, with if anything more comets of $i > 90^o$ rather than $i < 90^o$, although there are several comets included from the Kreutz Sun-grazing group at $i \sim 144^o$ — twelve in this plot, with more having been discovered since the catalogue of Marsden (1986) was compiled — which presumably represent one large comet which fragmented on some previous perihelion passage. If the LP comets represent a random influx from a spherically-symmetric Oort Cloud then a distribution varying as sin i would be expected: the plot seems to show this general trend although it is not possible with current statistics to prove whether the theoretical distribution is or is not followed by the data. It is worth pointing out that this in itself makes a difference of $\sim$ 50% in the calculated impact rate upon the Earth by LP comets (Olsson-Steel 1987d).

So can the LP comets supply the required meteoroid flux? Even if the production rate of meteoroids on elliptical orbits by these parent bodies was sufficient (Fulle 1987, 1988), there is still a major problem: that is, that the retrograde meteoroids are not in general on large orbits (see the plots given by Olsson-Steel & Lindblad (1989) for all available surveys, and by Jones & Sarma (1985) for the Canadian TV meteors). Many of these meteoroids have orbits with aphelia within 5 AU, and streams in retrograde Aten-like orbits have even been suggested (*e.g.* Terentjeva 1966; Sekanina 1973, 1976). Whilst one could suggest that these originated on large orbits and have since shrunk under the P–R effect, it would be difficult to explain how streams could remain as coherent entities over the necessary length of time, although retrograde orbits such as those are relatively stable against disruption in close planetary encounters (Olsson-Steel 1986, 1987e). We are thus forced to conclude that there must be a source of retrograde meteoroids which has a smaller mean orbit (*i.e.* the parent bodies are SP comets or AAA asteroids), and that the observed inclination distributions shown in Figure 3 are very much at variance with the 'real' distributions.

Could many retrograde SP objects (asteroids or comets) exist but escape discovery thus far in our exploration of the inner Solar System? We already have an example of a comet, namely P/Swift-Tuttle, which was a relatively bright object on its last observed passage within 1 AU, in the 1860's, and is also known to have undergone many active perihelion passages in the past since it is the progenitor of the well-developed Perseid meteoroid stream, but nevertheless was not observed on its expected return at the start of the 1980's. Although a later return is possible (Marsden 1973) it seems more likely that the comet has become temporarily dormant (*i.e.* it could have been recovered as a fast-moving asteroidal object). However, search techniques for near-Earth asteroids mitigate against the discovery of dark objects with such high angular velocities, so that the existence but non-discovery of such objects does not seem to be out of the question. Another example of the difficulty of finding fast-moving objects is the fact that it has only been in the last few years that any Jupiter-crossing asteroids have been found although the first (5025 P–L) was actually detected on plates exposed in the early 1960's (Olsson-Steel 1988b).

We therefore suggest that the meteor orbit data is *prima facie* evidence for the existence of a substantial population of planet-crossing asteroids and SP comets in high-inclination and retrograde orbits. Stagg & Bailey (1989) have made a similar suggestion. The confirmed existence of these would result in the calculated impact rate upon the Earth being severely altered, since the collision probability is a strong function of inclination (Olsson-Steel 1987d), and thus would be an important additional consideration in schemes for the obliteration of terrestrial fauna in global catastrophes.

References

Babadzhanov, P.B. & Obrubov, Yu.V. 1983. *Secular perturbations of Apollo, Amor and Aten asteroid orbits and theoretical radiants of meteor showers probably associated with them*, pp. 411-417. In

Streams, Publ. astron. Inst. Czechoslov. Acad. Sci., **67**, 141-150.

Bailey, M.E., Clube, S.V.M. & Napier, W.M. 1986. *The origin of comets*, Vistas Astron., **29**, 53-112.

Bailey, M.E., Wilkinson, D.A. & Wolfendale, A.W. 1987. *Can episodic comet showers explain the 30-Myr cyclicity in the terrestrial record?*, Mon.Not.Roy.Astron.Soc., **227**, 863-885.

Belyaev, N.A., Kresák, L., Pittich, E.M. & Pushkarev, A.N. 1986. CATALOGUE OF SHORT-PERIOD COMETS, Astronomical Institute of the Slovak Academy of Sciences, Bratislava.

Burns, J.A., Lamy, Ph. & Soter, S. 1979. *Radiative forces on small particles in the solar system*, Icarus, **40**, 1-48.

Carusi, A., Kresák, L., Perozzi, E. & Valsecchi, G.B. 1985. LONG-TERM EVOLUTION OF SHORT-PERIOD COMETS, Adam Hilger, Bristol.

Clube, S.V.M., 1987 *The origin of dust in the Solar System*, Phil. Trans. R. Soc. Lond. A **323**, 421-436.

Clube, S.V.M. & Napier, W.M. 1982. *Spiral arms, comets and terrestrial catastrophism*, Quart. Jl. Roy. Astron. Soc., **23**, 45-66.

Clube, S.V.M. & Napier, W.M. 1984. *The microstructure of terrestrial catastrophism*, Mon. Not. Roy. Astron. Soc., **211**, 953-968.

Clube, S.V.M. & Napier, W.M. 1986. *Giant comets and the galaxy: Implications of the terrestrial record.* In THE GALAXY AND THE SOLAR SYSTEM, eds. R. Smoluchowski, J.N. Bahcall & M.S. Matthews, University of Arizona Press, Tucson, pp. 260-295.

Cook, A.F. 1973. *A working list of meteor streams.* In EVOLUTIONARY AND PHYSICAL PROPERTIES OF METEOROIDS, eds. C.L. Hemenway, P.M. Millman & A.F. Cook, NASA SP-319, Washington, D.C., pp. 183-191.

Degewij, J. & Tedesco, E.F. 1982. *Do comets evolve into asteroids? Evidence from physical studies*, In COMETS, ed. L. Wilkening, University of Arizona Press, Tucson, pp. 665-695.

Dohnanyi, J. 1978. *Particle Dynamics*, In COSMIC DUST, ed. J.A.M. McDonnell, Wiley, Chichester, pp. 527-605.

Drummond, J.D. 1981. *A test of comet and meteor shower associations*, Icarus, **45**, 545-553.

Duncan, M., Quinn, T. & Tremaine, S. 1988. *The origin of short-period comets*, Astrophys. J., **328**, L69-L73.

Everhart, E. 1973. *The origin of short-period comets*, Astrophys. Lett., **10**, 131-135.

Fernández, J.A. 1985. *Dynamical capture and decay of short-period comets*, Icarus, **64**, 308-319.

Fulle, M. 1987. *Meteoroids from Comet Bennett 1970 II*, Astron. Astrophys. **183**, 392-396.

Fulle, M. 1987. *Meteoroids from Comet Bennett 1970 II*, Astron. Astrophys. **183**, 392-396.

Fulle, M. 1988. *Meteoroids from comets Arend-Roland 1957 III and Seki-Lines 1962 III*, Astron Astrophys., **189**, 281-291.

Gartrell, G. & Elford, W.G. 1975. *Southern hemisphere meteor stream determinations*, Aust. J. Phys., **28**, 591-620.

Grün, E., Zook, H.A., Fechtig, H. & Giese, R.H. 1985. *Collisional Balance of the Meteoritic Complex*, Icarus **62**, 244-272.

Gustafson, B.Å.S., Misconi, N.Y. & Rusk, E.T. 1987. *Interplanetary dust dynamics. III. Dust released from P/Encke: distribution with respect to the zodiacal cloud*, Icarus, **72**, 582-592.

Hahn, G. & Rickman, H. 1985. *Asteroids in cometary orbits*, Icarus, **61**, 417-442.

Hartmann, W.K., Tholen, D.J. & Cruikshank, D.P. 1987. *The relationship of active comets, 'extinct' comets, and dark asteroids*, Icarus, **69**, 33-50.

Hawkes, R.L., Jones, J. & Ceplecha, Z. 1984. *The Populations and Orbits of Double-Station TV Meteors*, Bull. Astron. Inst. Czechoslov., **35**, 46-64.

Hoffmeister, C. 1948. METEORSTRÖME, Verlag Johann Ambrosius Barth, Leipzig.

Hughes, D.W. 1978. *Meteors*, In COSMIC DUST, ed. J.A.M. McDonnell, Wiley, Chichester, pp. 123-185.

Hughes, D.W. 1985. *The transition between long-period comets, short-period comets and meteoroid streams*, In DYNAMICS OF COMETS: THEIR ORIGIN AND EVOLUTION, eds. A. Carusi & G.B. Valsecchi, Reidel, Dordrecht, pp. 129-142.

Hunt, J., Williams, I.P. & Fox, K. 1985. *Planetary perturbations on the Geminid meteor stream*, Mon. Not Roy. Astron. Soc., **217**, 533-538.

Hut, P., Alvarez, W., Elder, W.P., Hansen, T., Kauffman, E.G., Keller, G., Shoemaker, E.M. & Weissman, P.R. 1987. *Comet showers as a cause of mass extinctions*, Nature, **329**, 118-126.

Jones, J. & Sarma, T. 1985. *Double-Station Observations of 454 TV Meteors. II. Orbits*, Bull. Astron. Inst. Czechoslov., **36**, 103-115.

Kresák, L. 1978. *The Tunguska object: A fragment of comet Encke?* Bull. Astron. Inst. Czechoslov., **29**, 129-134.

Kresák, L. 1981. *The lifetimes and disappearance of periodic comets*, Bull. Astron. Inst. Czechoslov., **32**, 321-339.

Lebedinets, V.N., Korpusov, V.N. & Manokhina, A.V. 1981, 1982. RADIO METEOR INVESTIGATIONS IN OBNINSK: CATALOGUE OF ORBITS, VOLUMES 1 AND 2, Soviet Geophysical Committee of the Academy of Sciences of the U.S.S.R., Materials of the World Data Center B, Moscow.

Leinert, C., Röser, S. & Buitrago, J. 1983. *How to maintain the spatial distribution of interplanetary dust*, Astron. Astrophys., **118**, 345-357.

Marsden, B.G. 1973. *The next return of the comet of the Perseid meteors*, Astron. J., **78**, 654-662.

Marsden, B.G. 1986. CATALOGUE OF COMETARY ORBITS, I.A.U., Minor Planet Center, Cambridge, Mass.

McCrosky, R.E. & Posen, A. 1961. *Orbital elements of photographic meteors*, Smithson. Contrib. Astrophys., **4**, 15-84.

Napier, W.M. 1983. *The orbital evolution of short period comets*, In ASTEROIDS, COMETS, METEORS, eds. C.-L.Lagerkvist & H. Rickman, University of Uppsala Press, Sweden, pp. 391-395.

Nilsson, C.S. 1964. *A southern hemisphere radio survey of meteor streams*, Aust. J. Phys., **17**, 205-256.

Olsson-Steel, D. 1986. *The origin of the sporadic meteoroid component*, Mon. Not. Roy. Astron. Soc., **219**, 47-73.

Olsson-Steel, D. 1987a. *The dispersal of the Geminid meteoroid stream by radiative effects*, Mon. Not. Roy. Astron. Soc., **226**, 1-17.

Olsson-Steel, D. 1987b. *Asteroid 5025 P-L, comet 1967 II Rudnicki, and the Taurid meteoroid complex*, The Observatory, **107**, 157-160.

Olsson-Steel, D. 1987c. *Theoretical meteor radiants of Earth-approaching asteroids and comets*, Aust. J .Astron., **2**, 21-35.

Olsson-Steel, D. 1987d. *Collisions in the solar system -IV. Cometary impacts upon the planets*, Mon. Not. Roy. Astron. Soc., **227**, 501-524.

Olsson-Steel, D.I. 1987e. *The dynamical lifetime of comet P/Halley*, Astron. Astrophys., **187**, 909-912.

Olsson-Steel, D. 1988a. *Identification of Meteoroid Streams from Apollo Asteroids in the Adelaide Radar Orbit Surveys*, Icarus, **75**, 64-96.

Olsson-Steel, D.I. 1988b. *The dynamical lifetimes of Jupiter-crossing asteroids*, Astron. Astrophys., **204**, 313-316.

Olsson-Steel, D. 1988c. *The Taurid Complex and the Giant Comet Hypothesis*, The Observatory, **108**, 183-185.

Olsson-Steel, D. 1988d. *The inner Oort Cloud and the source of comet Halley*, Mon. Not. Roy. Astron. Soc., **234**, 389-399.

Olsson-Steel, D. & Elford, W.G. 1987. *The height distribution of radar meteors: Observations at 2 MHz*, J. Atmos. Terr. Phys., **49**, 243-258.

Olsson-Steel, D.I. & Lindblad, B.-A. 1989. *The Meteoroid Orbit Database*, Space Sci. Rev., (submitted).

Öpik, E.J. 1963. *The stray bodies in the solar system. I. Survival of cometary nuclei*, Adv. Astron. Astrophys., **2**, 219-262.

Porter, J.G. 1952. COMETS AND METEOR STREAMS, Chapman and Hall, London.

Rickman, H. 1985. *Interrelations between comets and asteroids*, In DYNAMICS OF COMETS: THEIR ORIGIN AND EVOLUTION, eds. A.

Carusi & G.B. Valsecchi, Reidel, Dordrecht, pp. 149-172.

Sekanina, Z. 1973. *Statistical model of meteor streams. III. Stream search among 19303 radio meteors*, Icarus, **18**, 253-284.

Sekanina, Z. 1976. *Statistical model of meteor streams. IV. A study of radio streams from the synoptic year*, Icarus, **27**, 265-321.

Singer, S.F. & Stanley, J.E. 1980. *Submicron particles in meteor streams.* In SOLID PARTICLES IN THE SOLAR SYSTEM, eds. I. Halliday & B.A. McIntosh, Reidel, Dordrecht, pp. 329-332.

Southworth, R.B. & Hawkins, G.S. 1963. *Statistics of meteor streams*, Smithson. Contrib. Astrophys., **7**, 261-285.

Stagg, C.R. & Bailey, M.E. 1989. *Stochastic capture of short-period comets*, Mon. Not. R. Aston. Soc., (in press).

Steel, D.I. & Elford, W.G. 1986. *Collisions in the Solar System - III. Meteoroid survival times*, Mon. Not. Roy. Astron. Soc., **218**, 185-199.

Štohl. J. 1986. *On meteor contribution by short-period comets*, PROC. 20TH ESLAB SYMP. ON THE EXPLORATION OF HALLEY'S COMET, ESA SP-250, Vol. II, 225-228.

Stothers, R.B. 1988. *Structure of Oort's Comet Cloud inferred from terrestrial impact craters*, The Observatory, **108**, 1-9.

Sykes, M.V. 1988. *IRAS observations of extended Zodiacal structures*, Astrophys. J., **334**, L55-L58.

Sykes, M.V. & Greenberg, R. 1986. *The formation and origin of the IRAS Zodiacal dust bands as a consequence of single collisions between asteroids*, Icarus, **65**, 51-69.

Terentjeva, A.K. 1966. *Minor meteor streams.* In RESULTS OF RESEARCHES OF INTERNATIONAL GEOPHYSICAL PROJECTS: METEOR INVESTIGATIONS, No. 1, Nauka, Moscow, pp.62-132.

Weissman, P.R. 1985. *Cometary Dynamics*, Space Sci. Rev., **41**, 299-349.

Wetherill, G.W. 1985. *Asteroidal source of ordinary chondrites*, Meteoritics, **20**, 1-22.

Whipple, F.L. 1967. *On maintaining the meteoritic complex*, In THE ZODIACAL LIGHT AND THE INTERPLANETARY MEDIUM, ed. J.L. Weinberg, NASA SP-150, Washington, DC, pp. 409-426.

Whipple, F.L. 1983. *1983 TB*, IAU Circ., No. 3881.

Williams, I.P. 1985. *The formation and evolution of meteor streams.* In DYNAMICS OF COMETS: THEIR ORIGIN AND EVOLUTION, eds. A. Carusi & G.B. Valsecchi, Reidel, Dordrecht, pp.115-127.

Williams, I.P., Johnson, C. & Fox, K. 1986. *Meteor Storms.* In ASTEROIDS, COMETS, METEORS II, eds. C.-I. Lagerkvist, B.-A. Lindblad, H. Lundstedt & H. Rickman, University of Uppsala Press, Sweden, pp. 559-563.

Wisdom, J. 1985. *Meteorites may follow a chaotic path to Earth*, Nature, **315**, 731-733.

Yeomans, D.K. 1981. *Comet Tempel-Tuttle and the Leonid meteors*, Icarus, **47**, 492-499.

Ziołkowski, K. 1988. *Revised orbit of 1967 II Rudnicki*, The Observatory, **108**, 182-183.

MAJOR WILDFIRES AT THE CRETACEOUS–TERTIARY BOUNDARY

Iain Gilmour, Wendy S. Wolbach and Edward Anders

Enrico Fermi Institute and Department of Chemistry, University of Chicago, Chicago, IL 60637-1433, USA.

Summary. K–T boundary clays from five sites are enriched in soot and charcoal by factors of 10^2–10^3 over Cretaceous levels, apparently due to a global fire. The soot profile nearly coincides with the Ir profile, implying that the fire was triggered by the impact. Much or all of the fuel was biomass, as indicated by the presence of retene (a hydrocarbon that is diagnostic of coniferous wood fires) and by the carbon isotopic composition (δC^{13}=25.4 ‰± 0.3 ‰), which resembles that of natural charcoal and atmospheric carbon particles from biomass fires. The amount of elemental C at the KTB (0.012 g cm^{-2}) is very large, and requires either that most of the Cretaceous biomass burned down or that the soot yield was higher than in small fires, *i.e.* > 2%.

At undisturbed sites, soot correlates tightly with Ir, As, Sb and Zn. Apparently soot and Ir-bearing ejecta particles — containing some volatile chalcophiles from the target rock — coagulated in the stratosphere, and then scavenged additional chalcophiles from the hydrosphere. In view of this coagulation, the K–T fire would only slightly prolong the period of darkness and cold caused by impact ejecta. However, it would contribute other environmental stresses, *e.g* a CO_2 greenhouse and a variety of pyrotoxins and mutagens. As the total of recognized

stresses has risen to 12, there is no basis for the contention that an impact cannot explain the observed selectivity of extinctions.

Introduction

The first hint of a fire at the K–T boundary was observed by Tschudy *et al.* (1984) at a Raton Basin site in Colorado. They found "large amounts of **fusinite**" in the basal coal layer containing the Ir anomaly and attributed it to "periods of fire consuming the vestigial or dead organic matter". Evidence for the worldwide nature of these fires was obtained by Wolbach *et al.* (1985), who found **soot** at KTB sites in Denmark, Spain and New Zealand, and attributed it to global fires of biomass (and perhaps fossil carbon) triggered by the impact.

To facilitate the discussion that follows, a few definitions:

Fusinite is carbonized woody material (i.e. natural **charcoal**) that is found in coal. It is ~ 10–100 μm in size and sometimes shows the cellular structure of wood.

Soot in the broad sense is the elemental carbon component of smoke, consisting of up to four subcomponents (Medalia & Rivin 1982):

Charcoal or coke particles lofted into the smoke plume.

Aciniform (grape-bunch-like) carbon, consisting of $10^{-2} - 10^{-1} \mu$m spherules of amorphous C that are welded into characteristic clusters and chains (Figure 1b); it is formed not by solid-phase charring of fuel but by gas-phase polymerization of C_2 radicals to icospirals (Kroto 1988).

Carbonaceous microgel, *ie* spheroidal carbon particles embedded in carbonaceous matrix.

Carbon cenospheres: 10–100 μm spheres formed by carbonization of liquid drops. As cenospheres are absent at the KTB and small charcoal particles are hard to distinguish from other carbonaceous materials of nondescript morphology, we use the term **soot** in a narrow sense, to include only the spheroidal components (2) and (3), above.

Kerogen is insoluble organic matter, formed by degradation of plant material. Depending on its source, **terrigenous** or **marine**, and thermal history, it can range from largely aliphatic to largely aromatic, with a parallel decrease in reactivity and H/C, O/C ratios. With increasing metamorphism, it grades continuously into **elemental carbon**, ranging from **amorphous carbon** to crystalline **graphite**. We shall use the generic term "carbon" or "elemental C" for the insoluble carbon remaining after removal of soluble minerals and reactive kerogen. It consists mainly of charcoal and soot, but with variable, usually small, amounts of resistant kerogen and detrital carbon from metamorphosed sediments.

Evidence for a global fire

Measurement of soot. The "insoluble carbon" remaining after acid dissolution of a sediment is usually dominated by kerogen, which prevents recognition, let alone measurement, of soot (Figures 1a,b). A

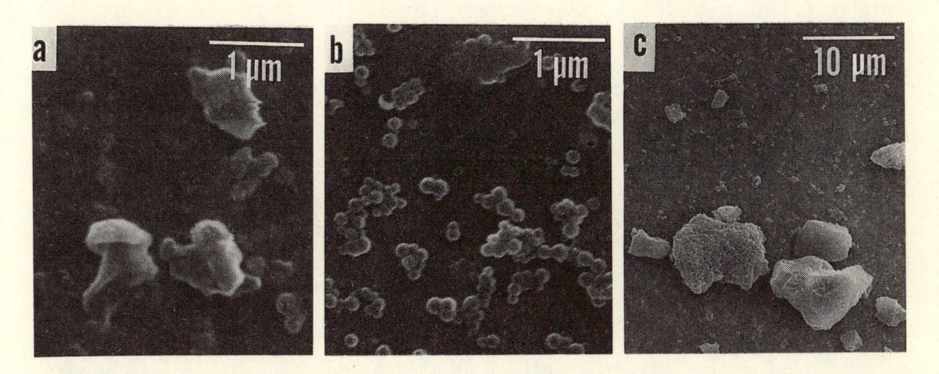

Figure 1. SEM photographs of demineralized carbonaceous fraction from K–T boundary clay at Woodside Creek, NZ. (a) Soot particles are barely discernible through the kerogen film. (b) After removal of kerogen by a 60 hour dichromate etch, soot spherules, with their characteristic "chained cluster" morphology, become clearly visible. (c) Coarse carbon of platy or pitted morphology. It is mainly charcoal in this sample, with minor amounts of resistant (terrigenous) kerogen and detrital, metamorphic carbon. Some detrital minerals — mainly rutile — are also present.:

perfect separation is impossible, since kerogen and elemental C form a structural continuum, with aromatic ring systems growing ever larger while H and other foreign atoms are progressively eliminated. The original extraction procedure improvised by Wolbach *et al.* (1985)

was too gentle, recovering all of the soot and other elemental C but leaving variable amounts of kerogen undissolved. Wolbach *et al.* (1988a) developed a harsher but more effective procedure involving dichromate oxidation under controlled conditions (Wolbach & Anders 1989). It destroys all the reactive kerogen and much of the resistant kerogen, while etching off only a small, easily correctable, fraction of the elemental C. The etched sample gives the abundance and δC^{13} of elemental C, whereas the unetched sample gives analogous data for kerogen (by difference). The mass fraction of soot and coarse carbon is determined by planimetric analysis on the SEM.

The "coarse carbon" has platy or pitted morphology (Figure 1c). It can include charcoal, detrital sedimentary carbon, and resistant (terrigenous) kerogen. Detrital carbon is hard to distinguish from charcoal, but usually is only a minor component. Biomass burning produces about 1-3$\times 10^{-4}$ g cm^{-2} yr^{-1} elemental C (Seiler & Crutzen 1980), compared to $\sim 2\times 10^{-5}$ g cm^{-2} yr^{-1} from weathering of sediments (Holland 1978). Moreover, the major part of the latter is oxidised to CO_2 rather than redeposited, as the C^{14} ages of modern sediments show no evidence of such "dead" carbon (Holland 1978). For this reason, essentially all elemental C in marine sediments is assumed to be charcoal (Goldberg 1985; Herring 1985). Resistant kerogen can be largely or entirely destroyed by a long, six hundred hour etch, but this has been done only for a few samples, mainly boundary clays. Thus our elemental C values, except for the boundary clay, may be too high by variable factors.

Carbon profiles across the K–T boundary. Figure 2 shows C and Ir profiles for the exceptionally well preserved Woodside Creek site, which has a very thin (0.6 cm) boundary clay layer with a very sharp Ir spike (Brooks *et al.* 1984; Wolbach *et al.* 1988a). Ir, elemental C and especially soot rise steeply at the boundary, by factors of 1400, 210 and 3600 respectively. The rise may be steeper than indicated in the graph, since both the -4.5 to 0 and -1.5 to 0 cm samples abut the boundary clay and may have been contaminated by slight mixing, imperfect sampling or Ir diffusion.

Similar enrichments of Ir, C and soot are found at Chancet Rocks (factors of 290, $\geq$ 790 and $\geq$ 660), and even at the severely burrowed and sheared Stevns Klint site, which shows a $\geq$ 180-fold increase in soot. Wolbach *et al.* (1985) had found only a 4-fold increase at the

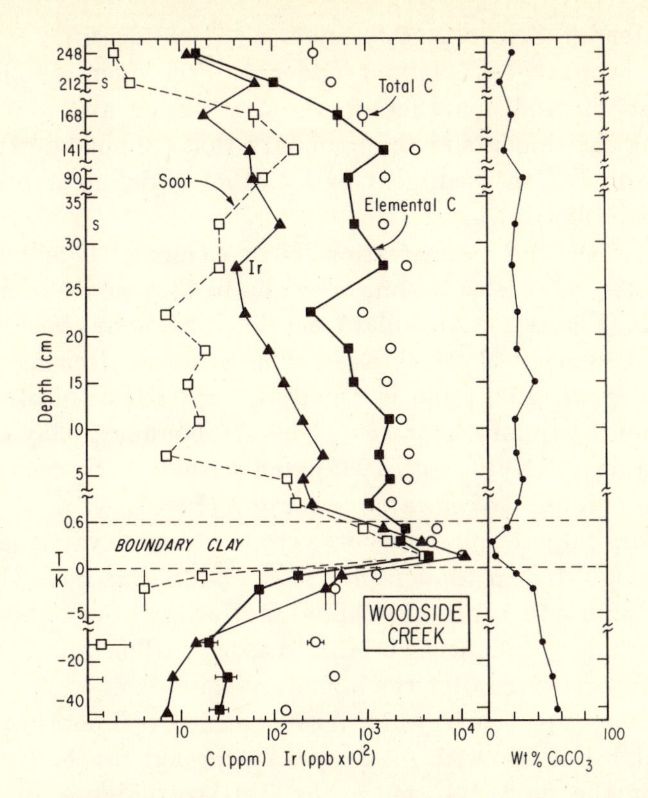

Figure 2. **Abundance profiles across the K-T boundary at Woodside Creek. Note expanded scales from −5 to +5 and especially 0 to 0.6 cm. Soot, Ir and elemental C all rise at the boundary by 2–3 orders of magnitude, and then slowly revert to Cretaceous levels. Soot appears at the basal 0.3 cm of the boundary clay, showing that the fires were in progress before the primary fallout had settled. (From Wolbach *et al* 1988a).:**

latter site (mainly due to incomplete removal of kerogen; see Wolbach *et al.* 1988a), but argued that the actual difference was as large as $\sim 10^3$ if one took into account the very short deposition time, less than a year, of the boundary clay. Opponents of the impact theory balk at accepting such short deposition times, apparently failing to realize that Stokes' law, which allows volcanic ash to settle in weeks or months (Ledbetter & Sparks 1979), also applies — with majestic impartiality — to impact ejecta, ensuring equally prompt fallout for a given particle size and water depth. Moreover, a detailed calculation based on the actual size distribution of quartz grains in boundary clay from Elendgraben in Austria gives an e-folding time of 0.58 – 0.75 yr for primary K–T fallout (Eder & Preisinger 1987). But even without this additional factor, the new carbon and soot values show striking enrichments at the boundary that cannot be explained by condensation of the sediment or by other enrichment processes (Anders *et al.* 1986).

In the Tertiary the concentrations of Ir, elemental C and soot all remain elevated to at least ~ 2m, reverting to Cretaceous levels only above 213 cm (Figure 2). A similar trend for Ir alone has been seen at other sites (Alvarez *et al.* 1980; Kastner *et al.* 1984; Preisinger *et al.* 1986; Strong *et al.* 1987) and is usually interpreted in terms of two kinds of fallout: primary, represented by the boundary clay or even only its basal layer (Kyte *et al.* 1985); and secondary, represented by the remaining material of elevated Ir content (Preisinger *et al.* 1986). This secondary fallout apparently was eroded from elevated sea-floor sites and redeposited in topographic lows. Such lows have the best chance of retaining their primary fallout, but at the price of becoming sinks for secondary fallout. For this reason, carbon and Ir values should be integrated only for the boundary clay.

Coagulation of soot with ejecta. At Woodside Creek soot correlates not only with Ir but also with As, Sb and Zn throughout the boundary clay and for the next 1.4 cm in the Tertiary (Figure 3). This correlation is very surprising in view of their separate origins: Ir from the meteorite, instantaneously; soot from fires, gradually over the next few weeks; As, Sb and Zn mainly adsorbed from seawater during final descent, perhaps with some contribution from the target rock (Gilmour & Anders 1989; Hildebrand & Wolbach 1989). The most likely explanation is that soot and Ir-bearing ejecta coagulated in the stratosphere, and then swept out As, Sb and Zn from ocean water, which had turned anoxic due to the mass death of plankton (Gilmour & Anders 1989). The initial coagulation may have been aided by the opposite electric charges of the soot and rock dust.

The soot/Ir ratio remains nearly constant for the first 6 cm of the Tertiary but then fluctuates widely about the mean (Figure 2; see also Wolbach *et al.* 1988b). Perhaps soot and Ir gradually became decoupled during lateral transport and redeposition, while the impact glass weathered to clay (Kastner *et al.* 1984).

Global distribution of carbon at the K-T boundary. Table 1

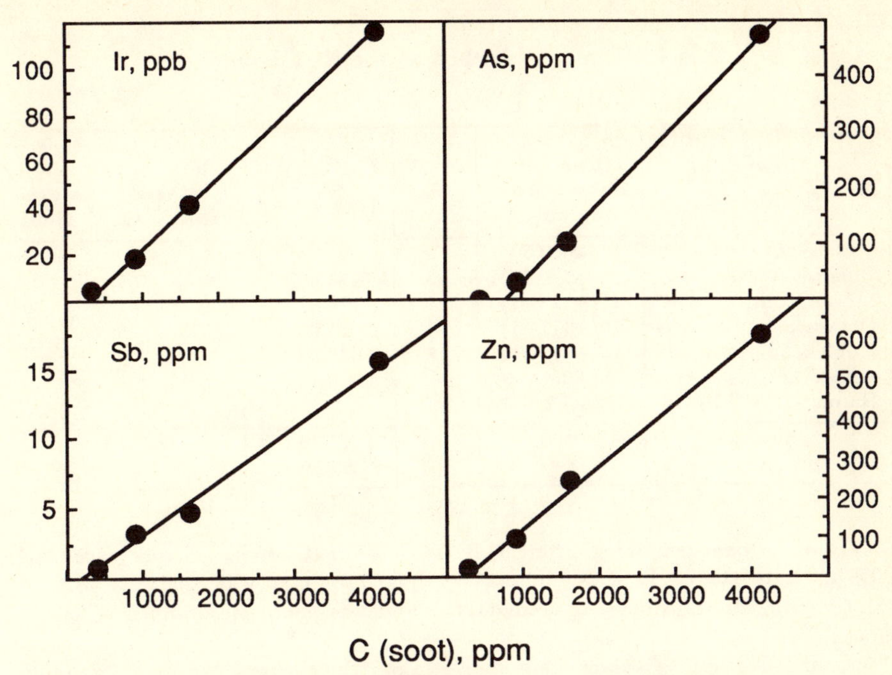

Figure 3. In the boundary clay and the next 3 cm at Woodside Creek, soot from biomass correlates tightly with Ir from the meteorite and As, Sb and Zn from terrestrial sources, despite their different origins. Presumably soot coagulated electrostatically with Ir-bearing impact ejecta in the stratosphere, and picked up As, Sb, Zn from vaporised rock, ocean water, or both:

summarizes carbon data for boundary clays at 7 sites. All show substantial amounts of elemental C but with variable proportions of soot. The first 5 sites contain 14–69% soot and have very similar δC^{13} values, with a mean of -25.4 ± 0.3‰. The last 2 sites have only 1–2% soot. Three other sites, for which data are incomplete, have high carbon but variable, often low, soot contents (Wolbach, in preparation).

On the basis of the first 5 sites, we have argued that charcoal and soot from the K–T fire rose into the stratosphere and became well mixed before falling back to Earth (Wolbach *et al.* 1988a). In view of the new data, various possibilities must be considered.

1. *The charcoal/soot ratio varied locally with fuel and combustion conditions, and may not have averaged out in the stratosphere if*

Table 1. Elemental carbon at the K–T boundary:

Site		Carbon abundance[a] mg cm^{-2}	Soot %	$\delta^{13}C_{PDB}$ ‰	Increase at K–T boundary over Cretaceous values Ir	C	Soot
Woodside Creek,	NZ	4.8±0.5	69	-25.23	1400	210	3600
Chancet Rocks,	NZ	35(+5/−20)	23	-25.42	290	≥790[b]	≥660[b]
Stevns Klint,	DK	11(+11/−1)	21	-25.81	2000	≥ 47[c]	≥180[c]
Caravaca,	E	10(+17/−1)	14	-25.00			
Gubbio,	I	13(+13/−1)	17	-25.48			
Agost,	E	3.8(+0.4/−0)	~2	–			
Raton (Goat Hill)	US	6.2(+3.2/−0)	~1	–			
Mean		12 ±4		-25.39 ±0.30			

[a] Total carbon remaining after 60 hour dichromate etch. Errors exceeding 20% indicate that only a portion of the boundary clay was analyzed or that the transition from primary to secondary fallout was poorly determined (Chancet Rocks).

[b] As the Cretaceous values were smaller than their errors, we used 2σ upper limits.

[c] Cretaceous marl from Nye Kløv, 190 km away, reanalyzed by Wolbach *et al.* (1988a). Tertiary limestone (+71 to +81 cm) from Stevns Klint gave similar values for C but a soot content of only ≤ 1 ppm, corresponding to a ≥ 2000-fold increase at the boundary.

mixing was limited. This is conceivable, but the variations would have to be quite large on a scale of a few hundred kilometres, to account for the differences among 5 Mediterranean sites (Table 1; *cf.* Wolbach, in preparation): three have 15–20% soot (Gubbio, Caravaca, El Kef) but two others, only a few hundred kilometres away, have only 1–2% (Agost, Hendaye).

2. *Unlike charcoal, soot never made it into the stratosphere but fell out locally, near its source.* Perhaps so, as only the smoke plumes from very intense fires extend into the stratosphere (Pittock *et al.* 1986, p. 110; hereafter SCOPE), but other things being equal, small soot particles should rise higher than large charcoal particles.

3. *The deposition of soot was controlled by rainout, and thus was highly localized, like present-day rainfall.* It is not clear, though, why this mechanism should selectively remove soot while permitting charcoal to spread globally.

4. *Soot and coarse carbon became separated at some sites by local currents.* Such a local mechanism is suggested by the variations among the Mediterranean sites. Anyhow, although these are interesting details that must be pursued and understood, they do not change the overall picture of a global charcoal-and-soot layer.

Sources of charcoal and soot

An asteroid or comet is not a plausible source. The total amount of C may be barely adequate, but it is present in combined, not elemental form. Conversion to soot in the fireball is at best inefficient and perhaps impossible, as the gas composition is fairly oxidizing (Wolbach *et al.* 1985), and any soot formed would combust as the fireball expands into the atmosphere. Moreover, δC^{13} of K–T carbon differs from that of meteoritic or cometary carbon, -25 vs -16 or +380 ‰(HCN in Comet Halley; Wyckoff *et al.* 1989), but is in the range of photosynthetic C, -20 to -30 ‰.

Biomass or fossil carbon? The δC^{13} values in Table 1 are remarkably constant and thus imply a single global source, either intrinsically uniform or homogenized by mixing. The mean of -25.4±0.3 ‰is suggestive of biomass source, since it resembles values for natural charcoal from forest fires (43% of the samples fall between -25 and -26 ‰; Deines 1980) and for carbon aerosols from biomass fires, -26.5±2 ‰(Cachier 1989), both of which are slightly fractionated relative to the parent plants. However, this is a weak argument, since at least some samples of fossil carbon fall in the same range (oil 4%, coal 18%).

A stronger argument comes from the polynuclear aromatic hydrocarbon or PAH, **retene** (1-methyl-7-isopropyl phenanthrene), which is present in boundary clay at Woodside Creek (Gilmour & Guenther 1988), Stevns Klint and Gubbio (Venkatesan & Dahl 1989), and at

DSDP site 605, off the east coast of the US (Simoneit & Beller 1987). This compound appears to be diagnostic of coniferous wood fires, and probably forms by low temperature pyrolysis of abietic acid, a common constituent of plant resins in conifers and some angiosperms (Ramdahl 1983; Gilmour & Guenther 1988). Apparently such trees provided at least part of the fuel for the K–T fire.

The boundary clay also is enriched up to a thousand-fold in various PAH's with 3–6 rings (Simoneit & Beller 1987; Gilmour & Guenther 1988; Venkatesan & Dahl 1989). Diagenesis can account for only part of the distribution; the remainder is apparently formed by combustion. Lacking alkyl side chains (in contrast to retene), these compounds evidently formed by high temperature synthesis from C_2 units rather than by gentle degradation of precursor molecules. Although not diagnostic of any particular fuel, these PAH's further strengthen the case for a major fire. The presence of the 6–ring PAH **coronene** (Gilmour & Guenther 1988; Venkatesan & Dahl 1989) implies conditions favouring growth of large ring systems, of which soot is the limiting case (a sheet of hexagons, curving onto itself due to the presence of $\sim$12 pentagons; Kroto 1988).

It is not not yet possible to estimate the fraction of resinous wood in the K-T fire. The retene/PAH ratio at Woodside Creek ($\sim$0.01-0.02) is smaller than that in air from Elverum in Norway (0.07), where half the total fuel is pine wood (Ramdahl 1983), and since resinous trees acounted for only a moderate fraction of the Maastrichtian biomass, these numbers are superficially consistent with a dominant biomass source. However, the retene/PAH ratio reflects not only the mass fraction but also the resin content of resinous wood and, more significantly, the conditions of combustion. Retene requires low temperatures and limited access of oxygen, and lacking such data for modern fires, let alone the K-T fire, we cannot draw any quantitative conclusions. We merely note that a major contribution of fossil C is not supported by the δC^{13} values or the absence of cenospheres.

Required amount of fuel. At first sight, biomass carbon would seem to be an ample source, as the amount of elemental C at the KTB, 0.012 g cm^{-2} (Table 1) is much smaller than the present above-ground biomass carbon of 0.2 g cm^{-2} (Seiler & Crutzen 1980), let alone the maximum Cretaceous value of 0.6 g cm^{-2} (estimated on the assumption that a land area equal to the present one had the same biomass density as present-day forests, namely 2 g cm^{-2}). However, this comparison is misleading; what matters is not the total biomass carbon A but the fraction F that was converted to *elemental C and dispersed as smoke* ("soot yield"). This fraction is

$$F = Af_b \varepsilon f_c$$

where f_b = the fraction of biomass carbon burned, ε = the fraction of f_b converted to smoke, and f_c = the fraction of elemental C in smoke.

The "baseline" parameters assumed for post-nuclear forest fires (National Research Council 1985; hereafter NRC) are: $f_b = 0.2$, $\varepsilon = 0.03$ and $f_c = 0.2$, with estimated uncertainty factors of 2-3, 2 and 3 respectively. The uncertainties in the latter two parameters are factors of 2 even in a single, controlled test fire (SCOPE, p.48). Using these values and A = 0.6 g cm^{-2}, we obtain F = 0.72 mg cm^{-2}, which is eighteen times less than the KTB value. Coincidentally, the combined uncertainties of the above parameters are twelve to eighteen times, but since these uncertainties are not independent, this does not eliminate the discrepancy. Either the soot yield was much higher in the K-T fire, or fossil carbon rather than biomass was the major fuel source.

Actually, there is a good reason for believing that the soot yield was much higher in the K-T fire. Seiler & Crutzen (1980) have pointed out that about one third of the "unburned" C in forest fires is converted to charcoal, *i.e.* 160 mg cm^{-2} in the above example. Less than 10% of this would need to be lofted to account for the 12 mg cm^{-2} of KTB carbon. Such coarse charcoal (10-20 μm) – which actually accounts for most of the carbon at the KTB (Table 1 and Figure 1) – was missed in test fires where f_c was determined from the light absorption (SCOPE, p. 51) but it has been detected in copious amounts by particle collections (NRC, p. 72).

Moreover, there are several profound differences between the test fires on which the NRC or SCOPE parameters are based and the KTB fire. First, test fires typically have dimensions of 1-100m, in contrast to 10^3-10^4 km for the KTB fire. It is known that ε and f_c increase for larger fires, as the air supply decreases and is more strongly preheated (NRC, pp.59, 62; SCOPE, pp.49, 52). Secondly, the impact and subsequent winds would strip the forest canopy, shatter heavy timbers, knock them down into the burning zone, and desiccate the fuel, thus raising the fuel density f_b, ε and f_c (NRC, pp. 49, 56, 59). Thirdly, smoke plumes and impact dust clouds would backscatter thermal radiation, likewise raising the above parameters (NRC, p. 56). Finally, the resulting larger and hotter fires would loft a larger proportion of the charcoal.

Given the very large potential reservoir of charcoal, some of which would reach the sea as river-borne runoff rather than atmospheric fallout, it appears that biomass alone may be an adequate source of the KTB carbon. However, fossil carbon remains a potential additional source. The amount of carbon excavated by the crater would not be large; a 150 km crater in *average* terrain with 2300 g cm^{-2} of carbon would eject 4 x 10^{17} g of C, or 0.08 g cm^{-2}, of which only an unknown, probably small, part would be converted to soot or charcoal in the relatively oxidizing medium of the fireball (Wolbach *et al.* 1985). Larger amounts might be produced by an impact into carbon-rich terrain, and by post-impact fires of oil seeps, exposed coal deposits or carbonaceous shales.

Ignition problem

Delayed fires. Living trees do not burn readily, and several authors have therefore proposed that the trees were first killed and freeze-dried, and ignited by lightning only after the sky had cleared and thunderstorms had resumed (Crutzen 1985; Argyle 1986; Schneider 1986). A prediction of this model is that carbon should overlie the Ir layer, but the data show no such trend: carbon appears even in the lowermost 0.3 cm of the boundary clay (Figure 2). Evidently the fires began well before the ejecta had settled.

High atmospheric O_2. Another alternative is that the O_2 content of the atmosphere was higher: at 24% O_2 by volume, living trees with $\sim$ 20% moisture have an ignition probability of $\sim$ 15%, compared to $<$ 1% at 21% O_2 (McMenamin & McMenamin 1987; Watson *et al.* 1978). For a more typical moisture content of 10%, these probabilities are 99% and 20%. However, there are three reasons why a higher O_2 content is not the answer.

First, it would be too much of a good thing; forests would burn down as fast as they grew. Indeed, Watson *et al.* (1978) contend that O_2 levels of 25–35% *"would be incompatible with land-based vegetation"*. Second, at present O_2 levels, only about .002 - .004 of land biomass burns down each year (Seiler & Crutzen 1980). This results in an aeolian carbon deposition rate of 3.9×10^{-6} g cm^{-2} yr^{-1} (Cachier *et al.* 1985), which in turn should yield mean carbon contents of 200–400 ppm in sediments of deposition rates of 10 to 5 cm over a thousand years, typical of Cretaceous shelf sediments. At higher O_2 contents a greater fraction of the land biomass would burn down each year, leading to correspondingly higher sedimentary carbon contents — provided soot yields remain the same, which is not certain. Anyhow, the observed elemental carbon contents in Cretaceous sediments are lower than 200–400 ppm, even though oxidative loss is ruled out by the presence of kerogen, thus: 18 ppm (Woodside Creek), $\leq$ 2ppm (Chancet Rocks), 3ppm (Nye Kløv) and $\leq$2ppm (Agost). Thirdly, the report of 30% O_2 in Cretaceous air trapped in amber (Berner & Landis 1988) has been questioned on a variety of grounds (Horibe & Craig 1988).

Ignition mechanisms. The most likely possibility is that trees were killed and dried by side effects of the impact, *e.g.* prompt heating of the atmosphere, the *"extraordinarily powerful wind, capable of flattening forests out to a distance of 500–1000 km"* which lasted for $\sim$ 1 hr and was followed by a very strong return wind (Emiliani *et al.* 1981), and by subsequent heating by the ejecta plume and the hot fallout (Öpik 1958). Those trees that were not ignited by the impact and its immediate after-effects might be burned down by a delayed process such as charge separation during settling of ejecta through the atmosphere, which, by analogy to volcanic ash falls (*eg* Perret 1935), should lead to extensive lightning activity, but on a global rather than a local scale. With multiple ignition points, fires would cover the globe in a matter of weeks, even if they spread at only the slow speed of normal wildfires, $\sim$ 5 km/hr.

Impact site(s). The prompt ignition radius for a 10^{31} erg impact has been estimated as 1200 km if the fuel is dry (O'Keefe & Ahrens 1988). A single impact thus would have to be on land or not too far offshore; for maximum burn area such an impact should be in the Bering Sea, as the fires would then spread across North America, Eurasia, and perhaps even Africa. But if a smaller area suffices (see above) then the impact site would merely need to be within $\sim 10^3$ km of a well-forested continent. Multiple impacts, as currently favoured, would require still less targeting. Finally, delayed ignition by fallout-induced lightning places no restrictions at all on impact site; even an impact in the deep ocean would cause fires on land. If this mechanism is important then all major impacts should cause fires.

Effects of fire

A fire producing 7×10^{16} g of soot would aggravate most of the environmental stresses of an impact, making a bad situation worse (Alvarez *et al.* 1980; Alvarez 1986; Wolbach *et al.* 1985, 1988a). These include:

Darkness and *cold*, would last longer, since soot absorbs sunlight more effectively than does rock dust (optical depth of 0.012 g cm^{-2} of soot is 1800). The cooling would be no greater, though, as it is already at the maximum.

Poisons such as NO and NO_2 (Lewis *et al.* 1982; Prinn & Fegley 1987) would be accompanied by $\sim$ 100 ppm CO and by a variety of organic pyrotoxins such as dioxins and PAH's.

Mutagens. Many pyrotoxins, especially PAH's, are mutagens, and may have caused the delayed extinctions of some Cretaceous "survivors", and may have speeded up the evolution of others.

Greenhouse effect. The $\sim$ 900 ppm of CO_2 expected to accompany the observed amount of soot (SCOPE p.217) would increase the greenhouse effect due to water vapor ($\sim$ 8°C, Emiliani *et al.* 1981) by another 5°to 10°C (Wolbach *et al.* 1986, Crutzen 1987), but it would last decades rather than months — depending on the rate of absorption by the oceans. The net temperature change of course also depends on other factors such as cloud cover and albedo (Rampino & Volk 1988) as well as persistence of the last remnants of dust and soot in the stratosphere.

Selectivity of extinction patterns

Some authors (Hickey 1981, Officer *et al.* 1987) have argued that

Table 2. Environmental stresses caused by the K–T impact:

Stress	Time Scale	Reference
Darkness	Months	ab
Cold	Months	abc
Winds (500 km/h)	Hours	d
Tsunamis	Hours	de
H_2O–Greenhouse	Months	d
CO_2–Greenhouse	Decades	fg
Fires	Months	hi
Pyrotoxins	Years	hijk
Acid Rain	Years	l
Destruction of Ozone Layer	Decades	lm
Impact-triggered Volcanism	Millenia?	n
Mutagens	Millenia	j

(a) Alvarez *et al.* (1980); (b) Toon *et al.* (1982); (c) Thompson (1988); (d) Emiliani *et al.* (1981); (e) Bourgeois *et al.* (1988); (f) Wolbach *et al.* (1986); (g) Crutzen *et al.* (1987); (h) Wolbach *et al.* (1985); (i) Wolbach *et al.* (1988a); (j) Gilmour & Guenther (1988); (k) Venkatesan & Dahl (1989); (l) Lewis *et al.* (1982); (m) Prinn & Fegley (1987); (n) Rampino & Stothers (1988).

the observed selectivity of the extinction patterns is inconsistent with the simple darkness-and-cold scenario of Alvarez *et al.* (1980). However, this scenario has been greatly extended in subsequent years by inclusion of other factors, exceeding the seven biblical plagues in number if not severity (Table 2). Virtually all of these twelve stresses are selective, inasmuch as they affect different species to different degrees. Some stresses pervade the entire Earth, whereas others vary with latitude, terrain, elevation, *etc.* And some taxa have hardy dormancy forms (spores, seeds, roots) or high reproduction ratios even for small breeding populations, whereas other do not. Anyhow, the selectivity problem may have been overstated; some authors contend that a cold spell alone could cause the observed pattern of plant extinctions (Wolfe & Upchurch 1986).

A first-order scientific task awaiting an imaginative ecologist is to rationalize the observed extinction patterns in terms of the above stresses. Some steps in this direction have already been taken, explaining the extinction of reptiles relative to mammals (Cowles 1939) or ammonites relative to nautiloids (Emiliani *et al.* 1981). Constructive work on this problem would be far more valuable than specious attacks on the impact theory.

Acknowledgements.
This work was supported in part by NSF grant EAR–8609218 and NASA fellowship NGT–50015. We are indebted to a Snowbird Conference participant for suggesting fallout-induced lightning as the ignition mechanism for forest fires.

References

Alvarez L., Alvarez W., Asaro F. & Michel H.V. 1980 *Extraterrestrial cause for the Cretaceous-Tertiary extinction.* Science **208**, 1095-1108.

Alvarez, W. 1986 *Toward a theory of impact crises.* EOS **67**, 649-658.

Argyle E. 1986 *Cretaceous extinctions and wildfires.* Science **234**, 261.

Anders E., Wolbach W.S. & Lewis R.S. 1986 *Cretaceous extinctions and wildfires.* Science **234**, 261-264.

Berner R.A. & Landis G.P. 1988 *Gas bubbles in fossil amber as possible indicators of the major gas composition of ancient air.* Science **239**, 1406-1409.

Bourgeois J., Hansen T.A., Wiberg P.L. & Kauffman E.G. 1988 *A tsunami deposit at the Cretaceous-Tertiary boundary in Texas.* Science **241**, 567-570.

Cachier H. 1989 *Isotopic characterization of carbonaceous aerosols.* Aerosol Sci. & Tech. in press.

Cachier H., Buat-Menard P., Fontugne M. & Rancher J. 1985 *Source terms and source strengths of the carbonaceous aerosol in the tropics.* J. Atmospheric Chem. **3**, 469-489.

Cowles R.B. 1939 *Possible implications of reptilian thermal tolerance.* Science **90**, 465-466.

Crutzen P.J. 1987 *Acid rain at the K/T boundary.* Nature **330**, 108-109.

Deines P. 1980 *The isotopic composition of reduced carbon in the terrestrial environment.* In HANDBOOK OF ENVIRONMENTAL ISOTOPE GEOCHEMISTRY (eds. P. Fritz & J. Ch. Fortes), pp. 329-407. Amsterdam: Elsevier.

Eder G. & Preisinger A. 1987 *Zeitstruktur globaler Ereignisse, veranschaulicht an der Kreide-Tertiär-Grenze.* Naturwissenschaften **74**, 35-37.

Emiliani C., Kraus E.B. & Shoemaker E.M. 1981 *Sudden death at the end of the Mesozoic.* Earth Planet. Sci. Lett. **55**, 317-334.

Gilmour I. & Anders E. 1989 *Cretaceous-Tertiary boundary event: Evidence for a short time scale.* Geochim. Cosmochim. Acta **53**, 503-511.

Gilmour I. &Guenther F. 1988 *The global Cretaceous-Tertiary fire: Biomass or fossil carbon?* In GLOBAL CATASTROPHES IN EARTH HISTORY. Snowbird, UT: Lunar Planet. Inst., pp. 60-61.

Goldberg E.D. 1985 BLACK CARBON IN THE ENVIRONMENT. New York: Wiley, 198 pp.

Herring J.R. 1985 *Charcoal fluxes into sediments of the North Pacific Ocean: The Cenozoic record of burning.* In THE CARBON CYCLE AND ATMOSPHERIC CO_2: NATURAL VARIATIONS ARCHEAN TO PRESENT. (eds. E.T. Sundquist & W.S. Broecker) pp. 419-442. Washington, D.C.: Amer. Geophys. Union.

Hickey L.J. 1981 *Land plant evidence compatible with gradual, not catastrophic, change at the end of the Cretaceous.* Nature **292**, 529-531.

Hildebrand A.R. & Wolbach W.S. 1989 *Carbon and chalcophiles at a nonmarine K/T boundary: Joint investigations of the Raton Section, NM.* Lunar Planet. Sci. **20, 414-415.**

Holland, H.D. 1978 THE CHEMISTRY OF THE ATMOSPHERE AND OCEANS, New York: Wiley-Interscience.

Horibe Y. & Craig H. 1988 *Is the air in amber ancient?* Science **241**, 720-721.

Kastner M., Asaro F., Michel H.V., Alvarez L.W. 1984 *The precursor of the Cretaceous-Tertiary boundary clays at Stevns Klint, Denmark, and DSDP Hole 465A.* Science **226**, 137-143.

Kroto H. 1988 *Space, stars, C_{60} and soot.* Science **242**, 1139-1145.

Kyte F.T., Smit J. & Wasson J.T. 1985 *Siderophile interelement variations in the Cretaceous-Tertiary boundary sediments from Caravaca, Spain.* Earth Planet. Sci. Lett. **73**, 183-195.

Ledbetter M.T. & Sparks R.S.J. 1979 *Duration of large-magnitude explosive eruptions deduced from graded bedding in deep-sea ash layers.* Geology, 7,240-244.

Lewis J.S., Watkins G.H. Hartman H. & Prinn R.G. 1982 *Consequences of major impact events on earth.* Geol. Soc. Am. Spec. Pap. **190**, 215-221.

McMenamin M.A.S. & McMenamin D.S. 1987 *Late Cretaceous atmospheric oxygen.* Science **235**, 1561-1562.

National Research Council 1985 THE EFFECTS ON THE ATMOSPHERE OF A MAJOR NUCLEAR EXCHANGE. Washington D.C.: National Academy Press, 193 pp.

Officer C.B., Hallam A., Drake C.L. & Devine J.D. 1987 *Late Cretaceous and paroxysmal Cretaceous/Tertiary extinctions.* Nature **326**, 143-149.

O'Keefe J.D., Ahrens T.J. & Koschny D. 1988 *Environmental effects of large impacts on the earth — relation to extinction mechanisms.* In GLOBAL CATASTROPHES IN EARTH HISTORY, Snowbird UT: Lunar Planet.Inst., pp. 133-134.

Öpik E.J. 1958 *On the catastrophic effects of collisions with celestial bodies.* Irish Astron. J. **5**, 34-36.

Perret F.A. 1935 THE ERUPTION OF MT. PELEE 1929–1932. Carnegie Institution of Washington, Pub. 458.

Pittock, A.B., Ackerman, T.A., Crutzen, P., MacCracken, M., Shapiro, C. & Turco, R.P. 1985 THE ENVIRONMENTAL CONSEQUENCES OF NUCLEAR WAR. Volume I: Physical. SCOPE 28a. Chichester: John Wiley & Sons.

Preisinger A., Zobetz E., Gratz A.J., Lahodynsky R., Becke M., Mauritsch H.J., Eder G., Grass F., Rogl F., Stradner H. & Surenian R. 1986 *The Cretaceous/Tertiary boundary in the Gosau Basin, Austria.* Nature **322**, 794-799.

Prinn R.G. & Fegley B. Jr. 1987 *Bolide impact, acid rain and biosphere traumas at the Cretaceous-Tertiary boundary.* Earth Planet. Sci. Lett. **83**, 1-15.

Ramdahl T. 1983 *Retene-a molecular marker of wood combustion in ambient air.* Nature **306**, 580-582.

Rampino M.R. & Stothers R.B. 1988 *Flood basalt volcanism during the past 250 million years.* Science **241**, 663-668.

Rampino M.R. & Volk T. 1988 *Mass extinctions, atmospheric sulphur and climatic warming at the K/T boundary.* Nature **332**, 63-65.

Schneider S.H., 1986 Unpublished manuscript.

Seiler W. & Crutzen P.J. 1980 *Estimates of gross and net fluxes of carbon between the biosphere and the atmosphere from biomass burning.* Climatic Change **2**, 207-247.

Simoneit B.R.T. & Beller H.R. 1987 *Lipid geochemistry of Cretaceous/Tertiary boundary sediments, Hole 605, Deep Sea Drilling Project Leg 93, and Stevns Klint, Denmark.* In INIT. DSDP (eds. J.E. van Hinte *et al.*) Vol. 93, Chap. 52, pp. 1211-1215. U.S. Govt. Printing Office.

Strong C.P., Brooks R.R., Wilson S.M., Reeves R.D., Orth C.J., Xue-Ying Mao, Quintana L.R. & Anders E. 1987 *A new Cretaceous-Tertiary boundary site at Flaxbourne River, New Zealand: Biostratigraphy and geochemistry.* Geochim. Cosmochim. Acta **51**, 2769-2777.

Thompson S.L. 1988 *Multi-year global climatic effects of atmospheric dust from large bolide impacts.* In GLOBAL CATASTROPHES IN EARTH HISTORY. Snowbird, UT: Lunar Planet. Inst., p. 194.

Toon O.B., Pollack J.B., Ackerman T.P., Turco R.P., McKay C.P. & Liu M.S. 1982 *Evolution of an impact generated dust cloud and its effects on the atmosphere.* Geol. Soc. Am. Spec. Pap. **190**, 187-200.

Tschudy R.H., Pillmore C.L., Orth C.J., Gilmore J.S. & Knight J.D. 1984 *Disruption of the terrestrial plant ecosystem at the Cretaceous-Tertiary boundary, Western Interior.* Science **225**, 1030-1032.

Venkatesan M.I. & Dahl J. 1989 *Further geochemical evidence for global fires at the Cretaceous-Tertiary boundary.* Nature, **338**, 57-60.

Watson A, Lovelock J.E. & Margulis L. 1978 *Methanogenesis, fires and the regulation of atmospheric oxygen.* BioSystems **10**, 293-298.

Wolbach W.S., Lewis R.S. & Anders E. 1985 *Cretaceous extinctions: Evidence for wildfires and search for meteoritic material.* Science **230**, 167-170.

Wolbach W.S., Lewis R.S., Anders E., Grady M.M., Pillinger C.T., Brooks R.R., Orth C.J. & Gilmore J.S. 1986 *Carbon isotopes and iridium at two Cretaceous-Tertiary (K–T) boundary sites in New Zealand.* Meteoritics **21**, 541-542.

Wolbach W.S., Gilmour I., Anders E., Orth C.J. & Brooks R.R. 1988a *A global fire at the Cretaceous-Tertiary boundary.* Nature **334**, 665-669.

Wolbach W.S., Anders E. & Orth C.J. 1988b *Darkness after the K–T impact: Effects of soot.* In GLOBAL CATASTROPHES IN EARTH HISTORY. Snowbird, UT: Lunar Planet. Inst., pp. 219-229.

Wolbach W.S. & Anders E. 1989 *Elemental C in sediments: Determination and isotopic analysis in the presence of kerogen.* Geochim. Cosmochim. Acta, **53**, in press.

Wolfe, J.A. & Upchurch G.R. Jr. 1986 *Vegetation, climatic and floral changes at the Cretaceous-Tertiary boundary.* Nature **324**, 148-152.

Wyckoff, S., Lindholm, E., Wehinger, P.A., Peterson, B.A., Zucconi, J.-M. & Festou, M.C. 1989 *The* $^{12}C/^{13}C$ *abundance ratio in Comet Halley.* Astrophys. J. **339**, 488-500.

Cometary dynamics: the inner core of the Oort cloud

M.E. Bailey

Department of Astronomy, The University, Manchester M13 9PL, U.K.

Summary. A brief introduction to the conventional Oort cloud theory of cometary origin is followed by a discussion of the evidence in favour of the cloud having a dense inner core. Recent research shows that the parameters of this inner core are severely constrained by the requirement that the cloud should not produce too many short-period comets. This argument complements similar constraints derived by others from the lack of strong surges in the terrestrial cratering record due to comet showers. If further work confirms the present short-period comet constraint on the inner core, comet showers probably have a negligible effect on the terrestrial cratering record, thereby further exacerbating the already severe difficulties in generating any 30 Myr cyclicity in the cratering record by plausible astronomical processes. Short-period comets are thus powerful probes of the inner core of the Oort cloud, and provide the exciting prospect that further dynamical and physical studies aimed at understanding their origin and evolution may eventually yield significant constraints on theories of cometary origin.

Introduction

The principal observational constraint which all theories of cometary origin have to satisfy is the underlying influx of long-period comets into the region of the terrestrial planets — the 'observed' near-parabolic flux. A large fraction of these comets have orbital periods well in excess of a million years, and they appear to come unpredictably from almost random directions in space at a rate on the order of one 'new' comet per year. Such comets, typically passing the sun at perihelion distances less than one or two astronomical units, suffer appreciable planetary perturbations during passage through the inner Solar System, with the result that their original orbits are significantly changed: either into a more tightly bound elliptical orbit or into an unbound hyperbolic type never to return. The near-parabolic flux therefore represents a net loss of comets, since those that end up in short-period orbits eventually decay, while those scattered into hyperbolic orbits are removed from the Solar System into interstellar space. The timescale for this cometary loss mechanism is a few orbital periods, leading to the removal of these long-period comets from the Solar System within times typically on the order of ten million years.

This line of argument means that comets are either a transient feature of Solar System evolution, having been recently 'captured' from interstellar space and now rapidly declining in numbers; or they truly represent a more or less steady-state primordial distribution and there is a substantial reservoir at large heliocentric distances to replenish those being lost. As is well known, Oort (1950) adopted the latter steady-state hypothesis, and was led to propose the existence of a huge, spherically symmetrical comet cloud surrounding the Sun and extending to large distances comparable in dimensions ($\approx 10^5$ AU) to those of the longest period new-comet orbits observed. The cloud was assumed to be approximately spherical in shape, so that the observed near-parabolic flux would appear roughly the same from all directions; while the huge extent of the cloud (with an outer radius on the order of a parsec) guaranteed that the comets moving in these regions would also be significantly affected by the perturbations of neighbouring, passing stars. Viewed from outside the Solar System — assuming that the comets could be seen — the general appearance of the Oort cloud would thus be rather like the spherically symmetrical image of a globular cluster, possibly resembling the picture of M 13 shown here as Figure 1.

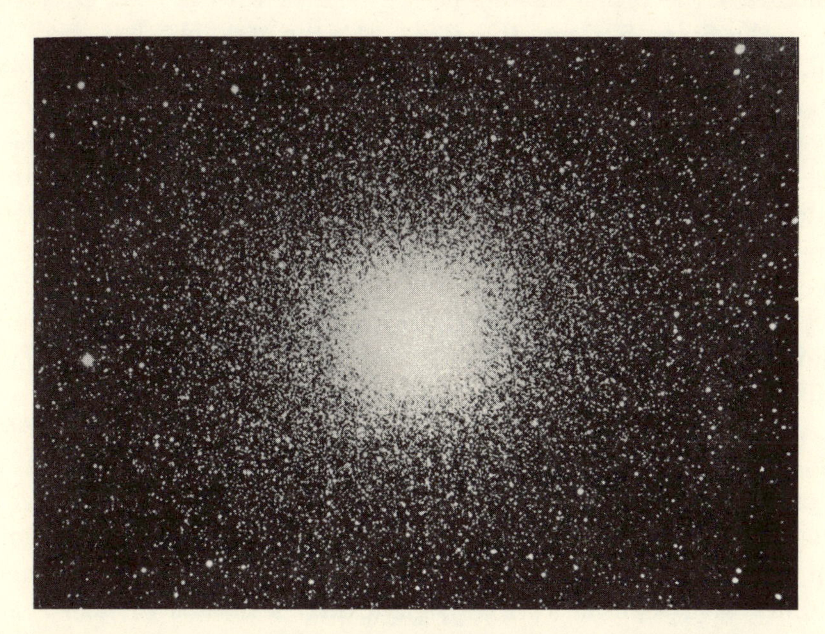

Figure 1. **Illustration of the globular cluster M 13 in Hercules, intended to give the general appearance of the Oort cloud of comets surrounding the Solar System.:**

The effect of stellar perturbations, and of the more recently recognized effects of the galactic tide (*e.g.* Byl 1986, Heisler & Tremaine 1986, Morris & Muller 1986, Torbett 1986a,b), is to ensure that long-period comets of sufficiently small perihelion distances to be observed are currently scattered into the inner Solar System at a more or less steady rate from the bulk of the comet cloud containing orbits of large semi-major axis and perihelia lying somewhat outside the planetary system, at least for original orbits with semi-major axes greater than a value a_t of order 3×10^4 AU (*cf.* Bailey 1983a, 1986a). This effectively explains the observed 'spike' in the frequency distribution of observed original $1/a$-values which originally formed the basis of Oort's hypothesis and has since been discussed by many authors. This is illustrated, for example, in Figure 2, which shows the observed $1/a$-distribution for all 225 well-observed long-period comets in the lists of Marsden, Sekanina & Everhart (1978) and Everhart &

Marsden (1983), taken from Bailey (1984).

Since stellar perturbations are expected to have almost completely randomized the orbits of long-period comets with $a > a_t$ over the age of the Solar System, the observed near-parabolic flux may be used to constrain the number density of comets in the cloud at great distance. For example, assuming the velocity distribution in the cloud is isotropic, a fraction $2q/a$ of comets with semi-major axis a will have perihelion distances less than q; and these all pass perihelion within a single orbital period $P(a) = 2\pi(GM_{\odot})^{-1/2}a^{3/2}$. Assuming an average semi-major axis of order $\bar{a} \simeq 7 \times 10^4$ AU for the comets in the Oort cloud and a corresponding near-parabolic flux of about one new comet per year within 3 AU (say), the required number of comets in the whole cloud is of order $N \simeq 2 \times 10^{11}$ (see Oort (1950) and Bailey (1983a) for a more precise determination of this quantity). The number density of comets in the cloud, assuming a mean heliocentric distance of order $1.5\bar{a}$, is then on the order of $3N/4\pi(1.5\bar{a})^3 \approx 4 \times 10^{-5}$ AU^{-3}. This is the order of magnitude of the cometary space density in the 'observed' part of the cloud, far from the Sun. The only remaining question, then, is to what extent is this value typical; and do there exist comets in orbits with semi-major axes significantly less than a_t, and if so, what is the distribution of *their* orbits?

Initially, Oort considered the possibility that the velocity distribution of the comets in the comet cloud would correspond to a Maxwellian distribution of relative velocities, thus leading (via the assumption of a steady-state model for the cloud as a whole) to a picture in which the density distribution increased inwards at such a great rate that the total number of comets in the cloud seemed to be physically unrealistic. (Any such density distribution has, of course, to be normalized to the 'observed' number density of the comets in the outer Oort cloud, so models in which the number of comets is assumed to be dominated by bodies relatively close to the planetary system in practice have also to be complemented by an assumed 'inner edge' to the cloud, in order not to conflict with general theoretical constraints on the maximum number of comets likely to be present in the inner Solar System or the neighbourhood of the outer planets.) Nevertheless, Oort found this sort of model unattractive from a theoretical point of view and subsequently abandoned it in favour of an alternative picture in which the number density of comets

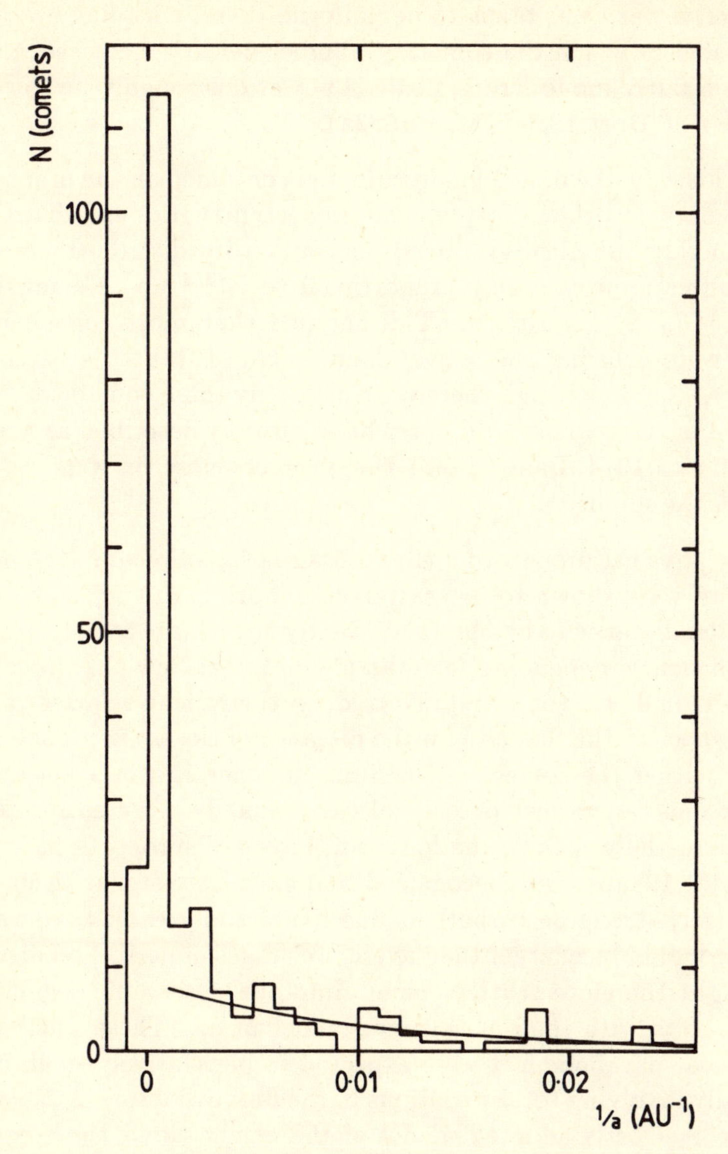

Figure 2. The observed frequency distribution of original $1/a$-values for 225 well-observed long-period comets. The solid curve represents an exponential fit to the data excluding the 'spike' near $1/a = 0$, corresponding to a mean fading (or decay) probability per revolution of order 2×10^{-3} (after Bailey 1984).:

in *velocity space* was taken to be uniform, thereby leading eventually to a model in which the cometary number density increased inwards at only a fairly modest rate, *i.e.* as a power law roughly proportional to $r^{-3/2}$ (*cf.* Oort 1950, Bailey 1983a).

In this way, the model finally adopted contained on the order of 2×10^{11} comets, in orbits of widely varying eccentricities and distributed with roughly spherical symmetry and a density distribution varying with radius approximately proportional to $r^{-3/2}$ up to a maximum distance $R_0 \simeq 2 \times 10^5$ AU. This implied that most comets in the model were orbiting the sun at heliocentric distances greater than 7×10^4 AU, and although there were definitely some comets at smaller distance (so the cloud could never be accurately described as a 'shell'; *cf.* Lyttleton 1974, Bailey 1977), these comets closer in were expected to be in the minority.

The physical process of stellar perturbations, by which comets in the cloud were shown to be scattered into orbits of sufficiently small perihelion distance to be observed, clearly has a large random element in its efficiency, depending for example on factors such as the mass of the individual star (or stars) involved, on their mean relative velocity with respect to the Sun and on the distance of closest approach of the stars to either the comets or the Sun. In general, over a long period of time one can expect occasional exceptionally close encounters to occur, and these will, in the long run, have a disproportionate effect (*cf.* Bailey 1986b). Oort recognized that such close stellar encounters, or similarly strong perturbations due to stars of greater than average mass, would induce larger than average changes in perihelion distance, and might therefore scatter comets into the observable region from initial orbits with semi-major axes less than a_t. These fluctuations in the near-parabolic flux were expected to be relatively small in size (typically involving total variations in the flux by factors $\lesssim 2$), largely because in Oort's adopted model of the comet cloud the comets of smaller semi-major axis were predicted to be in the minority. In recent years, however, a number of authors have begun to investigate alternative models for the density distribution in the cloud, most of these assuming that the majority of primordial comets now lie in otherwise undetectable orbits in the so-called 'inner core' of the Oort cloud.

Modern arguments

Oort's early discussion of the structure and dynamics of the primordial comet cloud, although largely anticipating much of the modern debate as to the mass and density distribution of the comets in the inner core of the cloud (and hence the possibility of significant variations in the near-parabolic flux: *i.e.* 'comet showers'), has since been overtaken by new arguments, emphasizing different aspects of Solar System dynamics and alternative views about the likely origin of comets (see Bailey, Clube & Napier 1986, 1989 for recent reviews). Öpik (1973), for example, constructed a model for the origin of comets based on the aggregation of dust grains in the neighbourhoods of the giant outer planets Jupiter, Saturn, Uranus and Neptune, and argued that the gravitational perturbations by these major planets would scatter the newly formed comets into orbits ranging up to of order 10^4 AU in size. Planetary perturbations were shown systematically to transfer energy to the cometary orbits, while stellar encounters gradually perturbed the comets of longest period into orbits of steadily increasing angular momentum, the overall effect being eventually to remove the comets from the zone of planetary influence at a characteristic time when their orbits first extended beyond about 5–10 $\times 10^3$ AU, *i.e. before* they had begun to populate the outer Oort cloud ($a \gtrsim 3 \times 10^4$ AU) in large numbers. With hindsight, this work can now be recognized as providing the first dynamical argument for the necessity, at least on some solar nebula theories of cometary origin, for the existence of a dense 'inner core' to the then conventional outer Oort cloud (*cf.* Fernández 1985, Duncan *et al.* 1987).

Later, Hills (1981) emphasized that the initial distribution of orbital energies assumed by Oort in his adopted model of the comet cloud (having most of the mass at large heliocentric distances) was not particularly plausible from a theoretical point of view, theoretical arguments based on comet formation within a generally collapsing solar nebula almost certainly predicting a final cometary density distribution having most of the mass closer in. He therefore strongly argued in favour of an alternative model in which the density increased sharply with decreasing radius, and considered a variety of examples with cometary number density scaling roughly as r^{-n}, with $n > 4$. Hills's paper subsequently proved extremely influential, since its main purpose was to highlight an important prediction of such strongly centrally condensed models of the Oort cloud: namely, the occurrence of occasional short-lived, but very intense, comet showers, as a result of exceptionally strong perturbations of cometary orbits within the inner core by close, slow-moving, or larger-than-average passing stars during the Sun's motion about the Galaxy. This work has now been extended, for example, by Fernández & Ip (1987), by Heisler, Tremaine & Alcock (1987) and by Hut *et al.* (1987).

Van den Bergh (1982) subsequently drew attention to the apparent necessity of a massive inner reservoir of comets within the conventional outer Oort cloud, in order to sustain the observed outer cloud population in the face of disruptive encounters by molecular clouds at intervals on the order of 5–10 $\times 10^8$ Myr, the effect of which had earlier been emphasized by Clube & Napier (1982) and Napier & Staniucha (1982) following the investigation by Napier & Clube (1979). At the same time, in work originally begun independently of both Hills and Van den Bergh, the present author (Bailey 1983a) drew these arguments together and showed not only that an inner core of a certain minimum density did indeed seem necessary for the survival of the observed 'outer' Oort cloud population, but also that it followed as an apparently natural consequence of generalized theoretical models of the cloud's structure and evolution under the effects of stellar and molecular cloud perturbations. In later papers Bailey (1983b,c,d) has also highlighted the potentially important implications of the inner core of the Oort cloud for a variety of investigations relating both to comets and Solar System dynamics, several of these ideas being described in more detail in a number of more recent investigations (*e.g.* Bailey 1986a,b,c).

Core constraints

Given the pivotal rôle now played by models of the inner core of the Oort cloud to theories of the origin of comets and of mechanisms by which such bodies might influence the Earth, it is important to develop arguments by which the parameters of the inner core might be constrained. Leaving aside infrared arguments (*cf.* Bailey *et al.* 1984, Jackson & Killen 1988) and those based on gravitational perturbations of the orbits of the outer planets (*e.g.* Whipple 1964, 1972; Bailey 1983d), which currently only strongly constrain those theories of the inner core which assume it to extend to within about a hundred astronomical units of the planetary system, the principal constraints obtained to date hinge either on cratering arguments or on the predicted number of short-period comets.

The terrestrial cratering constraint, originally outlined by Clube and Napier (Clube & Napier 1989, *cf.*Napier 1987) and by Stothers (1988), has been considered in more detail by Bailey & Stagg (1988). (The reader is referred to these papers for an introduction to the extensive literature dealing with the effects of cometary collisions with the Earth and the observed terrestrial cratering rate.) Here it seems that the principal uncertainty in coming to firm conclusions relates to the still uncertain masses and relative mass distribution of comets, and to the still controversial topic of the proportion of large craters produced either by active comets (or their decay products) or ordinary asteroids (ultimately presumed to come via collisions and subsequent gravitational perturbations from the asteroid belt). The crucial observation in this respect is that the terrestrial cratering rate during the past approximately 250 Myr seems not to have varied by much more than a factor of two or so, whilst on a much longer timescale it is consistent with a roughly constant (or weakly increasing) rate over the past 3×10^9 years. This is determined, for example, by comparisons with the lunar cratering record, which also indicates a roughly constant or weakly increasing rate of production of large craters (Baldwin 1985).

If comets are indeed the major contributor to the terrestrial and lunar impactors, these cratering constraints may provide strong constraints on the parameters of the inner core. For example, a primordial inner core of the Oort cloud must on general evolutionary grounds not only be a steadily declining population, but must also be one which can replenish the outer Oort cloud for the entire age of the Solar System without sensible change and without at the same time being so dense that the very large variations in the cometary flux discussed by Hills (1981) become the norm rather than the exception. Napier (1987) and Stothers (1988) have argued that these constraints are so severe as almost to rule out the dense inner core hypothesis, although Bailey & Stagg (1988) have emphasized that the cratering constraints may be significantly relaxed by taking what appears to be a more self-consistent (albeit still uncertain) distribution of cometary masses, and allowing for the possibly dominant contribution to terrestrial craters coming from bodies appearing to originate in the asteroid belt rather than coming directly from the Oort cloud.

None of these arguments, of course, can yet be considered as the last word in these matters, and it is important to develop more detailed and internally self-consistent models of the inner core, taking due consideration not only of the most recent calibrations of cometary masses and their influx into the planetary system, but also of the time-dependent evolution of the inner core and independent constraints on its properties provided by observations of the craters on other planets and planetary satellites (*e.g.* Shoemaker & Wolfe 1982). In this way, the cratering constraints on the inner core of the Oort cloud will eventually become much more precise.

A second important constraint on the inner core — the short-period comet constraint — was also emphasized by Bailey (1986c) who, following other authors (*cf.* Bailey 1983b and references therein), pointed out that one consequence of a sufficiently massive inner core of the Oort cloud would be the resolution of the long-standing 'short-period comet problem', namely the difficulty of explaining the observed number of short-period comets as a simple consequence of gravitational capture by planetary perturbations of comets comprising the near-parabolic flux from the outer Oort cloud. The main point to be emphasized is that although the probability of long-period comets eventually evolving into short-period orbits is a decreasing function of the comets' initial perihelion distance (mainly due to the smaller masses of the outer planets Uranus and Neptune compared to Jupiter and Saturn), the predicted flux of long-period comets in the outer solar system should be very much greater in the Uranus-Neptune zone than it is in the region of the terrestrial planets. This is because these outer planets lie outside the 'loss-cone' of the cometary velocity distribution determined by the larger perturbations of Jupiter and Saturn, allowing them to sample the intrinsic, virtually undiminished near-parabolic flux from the whole of the Oort cloud, rather than just that part of the flux corresponding to comets with semi-major axes $a \gtrsim a_t \simeq 3 \times 10^4\,\mathrm{AU}$. The result (*cf.* Hills 1981) is that the outer planets experience a net cometary bombardment equivalent in order of magnitude to the most intense comet shower experienced in the neighbourhood of the Earth, virtually the whole time.

Thus, if the inner core has a sufficient mass or degree of central concentration, this greater influx of comets with perihelia beyond about 15 AU can more than compensate for the lower net capture probability of such comets into eventually observable short-period orbits, thereby allowing such comets to dominate the supply of short-period comets, and so potentially resolve the short-period comet problem. However, the inner core must not be so massive as to produce *too many* short-period comets, and the argument therefore provides an important indirect constraint on the parameters of the assumed inner core of the Oort cloud.

Recent work by Stagg & Bailey (1989), following Everhart (1977), has determined the mean capture probability for original long-period orbits with perihelia in the Uranus-Neptune zone and inclinations less than 27°. Their results, together with the observed number of short-period comets, may thus be used to provide a rough constraint on the cometary influx through the outer planetary system. Modelling the inner core with an 'inner edge' corresponding to a semi-major axis $a_0 \approx 10^3$–10^4 AU and a power-law distribution of orbital energies with index γ to be determined (so the total number of comets with semi-major axes in the range $[a, a + da]$ is assumed to be proportional to $a^{\gamma-2}$), the derived constraint reduces to the approximate relation

$$\left(\frac{a_0}{a_t}\right)^{\gamma-7/2} \approx 10^3$$

where $a_t \simeq 3 \times 10^4$ AU.

It is remarkable that this constraint gives a result for the parameters of the inner core which lies close to the predictions of the planetesimal hypothesis (*e.g.* Öpik 1973, Fernández 1985, Duncan *et al.* 1987), namely an inner edge corresponding to a semi-major axis of order $a_0 \simeq (3000, 4000, 6500)$ AU for $\gamma = (0.5, 0.0, -1.0)$ respectively. The inferred maximum intensity of comet showers from such an inner core then implies a maximum variation in the terrestrial cratering rate within the limit apparently allowed by observations; while it is important to note that such an inner core may also be sufficient to replenish the outer Oort cloud over the age of the Solar System.

If, however, a substantial number of short-period comets come from the comet belt recently reintroduced by Duncan *et al.* (1988) in order to explain the flattened distribution of inclinations of the Jupiter-family short-period comets (*cf.* Edgeworth 1949, Kuiper 1951, Whipple 1964, Mendis 1973, Fernández 1980), the constraint on any spherically symmetrical inner extension of the Oort cloud would become much tighter. Reducing the number of observed short-period comets to be explained as a result of capture from the long-period flux emanating from the Oort cloud (corresponding to the assumption that the Jupiter-family comets come from a comet belt or some other source, such as the break-up of an incoming 'giant' comet), would effectively rule out any very dense model of the inner core, with the consequent difficulty of accommodating large variations in the cometary flux within such a model. The picture of the inner core of the Oort cloud would then reduce to that envisaged by Stothers (1988), or even that of Oort (1950); but the core might then be insufficiently massive to explain the survival of the outer 'observed' Oort cloud for the age of the solar system!

Conclusion

Recent models of the Oort cloud which postulate a massive inward extension of the observed outer cometary system (a dense inner core hypothesis) are strongly constrained by both the terrestrial cratering record (which shows no evidence for strong surges in the cratering rate) and the observed number of short-period comets. Such constraints, combined with knowledge of the Sun's galactic environment as currently understood, make the likelihood of significant 30 Myr variations in the cometary flux about the Earth due to a fluctuating sequence of showers extremely remote (*cf.* Bailey, Wilkinson & Wolfendale 1987; Wolfendale & Wilkinson, these proceedings). Indeed, models in which significant variations in the terrestrial cratering rate are attributed to quasi-periodic comet showers, arising as a result of stellar (or other) perturbations on a dense inner core of the Oort cloud, need to be checked for consistency with the observed number of short-period comets predicted also to come from the postulated dense inner core.

In this way the origin of short-period comets, whether from a comet belt lying just beyond the planetary system or a more spherically symmetrical inner core of the Oort cloud, emerges as a central problem in discussions of cometary cosmogony and of the interrelationship with theories of terrestrial catastrophism. Whereas the observed near-parabolic flux has long provided constraints on the properties of the outer Oort cloud, short-period comets are now increasingly recognized as important probes of the hypothetical inner core. In this way, short-period comets may eventually provide the key to understanding not only the origin of comets generally, but also the evolution and origin of the Oort cloud itself.

Acknowledgments

This work was supported by the UK Science and Engineering Research Council.

References

Bailey, M.E., 1977. *Some comments on the Oort Cloud*, Astrophys. Space Sci., **50**, 3–22.

Bailey, M.E., 1983a. *The structure and evolution of the Solar System comet cloud*, Mon. Not. R. Astron. Soc., **204**, 603–633.

Bailey, M.E., 1983b. *Is there a dense primordial cloud of comets just beyond Pluto?* Asteroids Comets Meteors, eds. Lagerkvist, C.-I. & Rickman, H., 383–386. Uppsala Observatory, Uppsala, Sweden.

Bailey, M.E., 1983c. *Theories of cometary origin and the brightness of the infrared sky*, Mon. Not. R. Astron. Soc., **205**, 47P–52P.

Bailey, M.E., 1983d. *Comets, Planet X, and the orbit of Neptune*, Nature, **302**, 399–400.

Bailey, M.E., 1984. *The steady-state 1/a-distribution and the problem of cometary fading*, Mon. Not. R. Astron. Soc., **211**, 347–368.

Bailey, M.E., 1986a. *The mean energy transfer rate to comets in the Oort Cloud and implications for cometary origins*, Mon. Not. R. Astron. Soc., **218**, 1–30.

Bailey, M.E., 1986b. *A note on the mean energy transfer rate by point-mass perturbers*, ASTEROIDS COMETS METEORS II, eds, Lagerkvist, C.-I., Lindblad, B.A., Lundstedt, H. & Rickman, H., 207–210. Uppsala University, Uppsala, Sweden.

Bailey, M.E., 1986c. *The near-parabolic flux and the origin of short-period comets,* Nature, **324,** 350–352.

Bailey, M.E., Clube, S.V.M. & Napier, W.M., 1986. *The origin of comets,* Vistas Astron., **29,** 53–112.

Bailey, M.E., Clube, S.V.M. & Napier, W.M., 1989. THE ORIGIN OF COMETS. Pergamon Press, Oxford.

Bailey, M.E., McBreen, B. & Ray, T.P., 1984. *Constraints on cometary origin from the isotropy of the microwave background and other measurements,* Mon. Not. R. Astron. Soc., **209,** 881–890. *Corrigendum:* **211,** 255.

Bailey, M.E. & Stagg, C.R., 1988. *Cratering constraints on the inner Oort cloud: steady-state models,* Mon. Not. R. Astron. Soc., **235,** 1–32.

Bailey, M.E., Wilkinson, D.A. & Wolfendale, A.W., 1987. *Can episodic comet showers explain the 30-Myr cyclicity in the terrestrial record?* Mon. Not. R. Astron. Soc., **227,** 863–885.

Baldwin, R.B., 1985. *Relative and absolute ages of individual craters and the rate of infalls on the Moon in the post-Imbrium period,* Icarus, **61,** 63–91.

Byl, J., 1986. *The effect of the Galaxy on cometary orbits,* Earth, Moon, Planets, **36,** 263–273.

Clube, S.V.M. & Napier, W.M., 1982. *Spiral arms, comets and terrestrial catastrophism,* Q. Jl. R. Astron. Soc., **23,** 45–66.

Clube, S.V.M. & Napier, W.M., 1989. *An episodic-cum-periodic galacto-terrestrial relationship,* Q. Jl. R. Astron. Soc., submitted.

Duncan, M., Quinn, T. & Tremaine, S.D., 1987. *The formation and extent of the solar system comet cloud,* Astron. J., **94,** 1330–1338.

Duncan, M., Quinn, T. & Tremaine, S.D., 1988. *The origin of short-period comets,* Astrophys. J. Lett., **328,** L69–L73.

Edgeworth, K.E., 1949. *The origin and evolution of the solar system,* Mon. Not. R. Astron. Soc., **109,** 600–609.

Everhart, E., 1977. *Evolution of comet orbits as perturbed by Uranus and Neptune,* COMETS ASTEROIDS METEORITES: INTERRELATIONS, EVOLUTION AND ORIGINS, ed. Delsemme, A.H., IAU Coll. No. 39, 99–104. University of Toledo, Toledo, Ohio.

Everhart, E. & Marsden, B.G., 1983. *New original and future cometary orbits,* Astron. J., **88,** 135–137.

Fernández, J.A., 1980. *On the existence of a comet belt beyond Neptune,* Mon. Not. R. Astron. Soc., **192,** 481–491.

Fernández, J.A., 1985. *The formation and dynamical survival of the Oort Cloud,* DYNAMICS OF COMETS: THEIR ORIGIN AND EVOLUTION, eds. Carusi, A. & Valsecchi, G.B., IAU Coll. No. 83, 45–70. (Astrophys. Space Sci. Lib. **115.**) Reidel, Dordrecht, The Netherlands.

Fernández, J.A. & Ip, W.-H., 1987. *Time-dependent injection of Oort-cloud comets into Earth-crossing orbits*, Icarus, **71**, 46–56.

Heisler, J. & Tremaine, S.D., 1986. *The influence of the galactic tidal field on the Oort comet cloud*, Icarus, **65**, 13–26.

Heisler, J., Tremaine, S.D. & Alcock, C., 1987. *The frequency and intensity of comet showers from the Oort cloud*, Icarus, **70**, 269–288.

Hills, J.G., 1981. *Comet showers and the steady-state infall of comets from the Oort Cloud*, Astron. J., **86**, 1730–1740.

Hut, P., Alvarez, W., Elder, W.P., Hansen, T., Kauffman, E.G., Keller, G., Shoemaker, E.M. & Weissman, P.R., 1987. *Comet showers as a cause of mass extinctions*, Nature, **329**, 118–126.

Jackson, A.A. & Killen, R.M., 1988. *Infrared brightness of a comet belt beyond Neptune*, Earth, Moon, Planets, **42**, 41–47.

Kuiper, G.P., 1951. *On the origin of the Solar System*, ASTROPHYSICS, ed. Hynek, J.A., 357–424. McGraw-Hill, New York.

Lyttleton, R.A., 1974. *The non-existence of the Oort cometary shell*, Astrophys. Space Sci., **31**, 385–401.

Marsden, B.G., Sekanina, Z. & Everhart, E., 1978. *New osculating orbits for 110 comets and analysis of original orbits for 200 comets*, Astron. J., **83**, 64–71.

Mendis, D.A., 1973. *The comet – meteor stream complex*, Astrophys. Space Sci., **20**, 165–176.

Morris, D.E. & Muller, R.A., 1986. *Tidal gravitational forces: the infall of "new" comets and comet showers*, Icarus, **65**, 1–12.

Napier, W.M., 1987. *The origin and evolution of the Oort cloud*, Interplanetary Matter, eds. Ceplecha, Z. & Pecina, P., PROC. TENTH EUROPEAN REGIONAL MEETING IN ASTRONOMY, VOL. 2, PRAGUE, 13–19.

Napier, W.M. & Clube, S.V.M., 1979. *A theory of terrestrial catastrophism*, Nature, **282**, 455–459.

Napier, W.M. & Staniucha, M., 1982. *Interstellar planetesimals — I. Dissipation of a primordial cloud of comets by tidal encounters with nebulae*, Mon. Not. R. Astron. Soc., **198**, 723–735.

Oort, J.H., 1950. *The structure of the cloud of comets surrounding the solar system and a hypothesis concerning its origin*, Bull. Astron. Inst. Neth., **11**, 91–110.

Öpik, E.J., 1973. *Comets and the formation of planets*, Astrophys. Space Sci., **21**, 307–398.

Shoemaker, E.M. & Wolfe, R.F., 1982. *Cratering time scales for the Galilean satellites*, SATELLITES OF JUPITER, ed. Morrison, D., 277–339. University of Arizona Press, Tucson, USA.

Stagg, C.R. & Bailey, M.E., 1989. *Stochastic capture of short-period comets*, Mon. Not. R. Astron. Soc., in press.

Stothers, R.B., 1988. *Structure of Oort's comet cloud inferred from terrestrial impact craters*, Observatory, **108**, 1–9.

Torbett, M.V., 1986a. *Dynamical influence of galactic tides and molecular clouds on the Oort cloud of comets*, THE GALAXY AND THE SOLAR SYSTEM, eds. Smoluchowski, R., Bahcall, J.N. & Matthews, M.S., 147–172. University of Arizona Press, Tucson.

Torbett, M.V., 1986b. *Injection of Oort cloud comets to the inner solar system by galactic tidal fields*, Mon. Not. R. Astron. Soc., **223**, 885–895.

Whipple, F.L., 1964. *The evidence for a comet belt beyond Neptune*, Proc. Natl. Acad. Sci. (USA), **51**, 711–718.

Whipple, F.L., 1972. *The origin of comets*, THE MOTION, EVOLUTION OF ORBITS, AND ORIGIN OF COMETS, eds. Chebotarev, G.A., Kazimirchak-Polonskaya, E.I. & Marsden, B.G., IAU Symp. No. 45, 401–408. Reidel, Dordrecht, The Netherlands.

PERIODIC MASS EXTINCTIONS: SOME ASTRONOMICAL DIFFICULTIES

A.W. Wolfendale and D.A. Wilkinson

Department of Physics, University of Durham, South Road, Durham DH1 3LE, U.K.

Summary. Although it is almost certain that comets are shaken free from the Oort Cloud by encounters with stars and molecular clouds we have found no process involving these or other perturbers which would give rise to a 30 Myr periodicity (or any other periodicity) in their rate of arrival at the Earth.

Introduction

The idea of cometary impacts on the Earth being responsible for mass extinctions of organisms has become respectable in recent years largely through the work of Clube (1978), Napier & Clube (1979), and Alvarez *et al.* (1980). The last mentioned work, and later contributions by the same authors, has been singularly important in that it has identified an iridium layer in clay deposits at an age of about 65 Myr BP, the iridium concentration being suggestive of cometary or asteroidal impact. At this time, 65 Myr BP, there was the well known 'death of the dinosaurs' and the date corresponds to the geological Cretaceous-Tertiary boundary. Thus, there is circumstantial evidence for a comet (or comets) causing this particular extinction.

Another landmark in the subject was the claim by Raup & Sepkoski (1984) that there is a regular 26 Myr periodicity in the pattern of extinctions. Figure 1 from an earlier paper (Bailey *et al.* 1987) shows the situation. Insofar as the 26 Myr period is roughly the

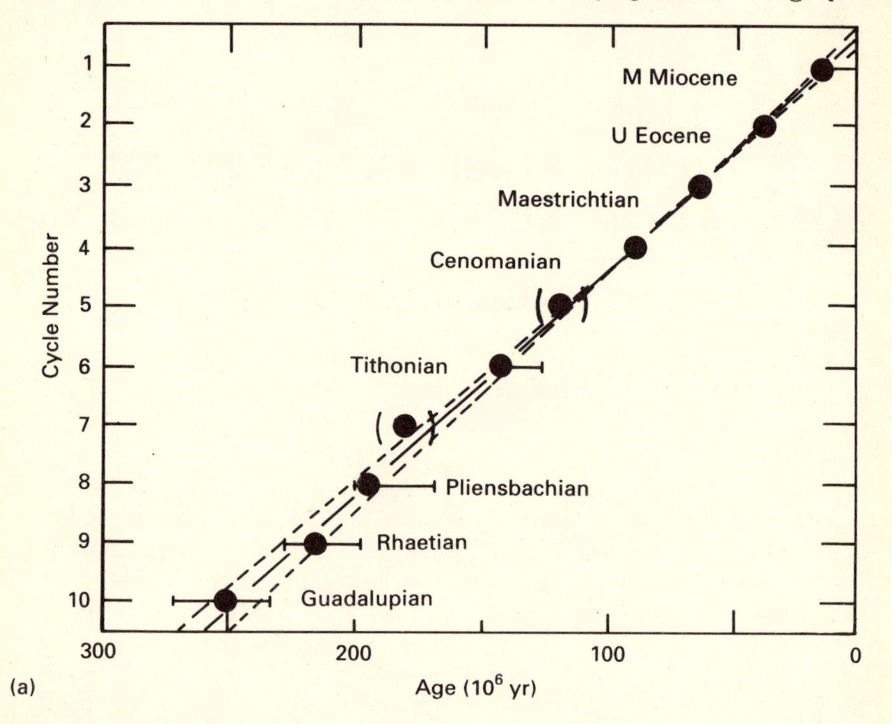

Figure 1(a). The cycle number versus age relationship is shown for eight significant extinction events (Sepkoski and Raup 1986). The solid line is a least-squares fit with slope corresponding to a period of 26.2 Myr, while the dashed lines show the range in slopes for alternative solutions which pass through the error bars of all the points.:

time interval between successive crossings of the Galactic plane by the Solar System in its motion around the Galaxy, an astronomical origin for the extinctions appears to be strengthened. An increased flux of comets coincident with Galactic plane crossings is a natural hypothesis and such an increase could in principle be due to interactions of giant molecular clouds (GMCs) with the Oort cloud whereby heightened fluxes of comets in the inner planetary system would result (Rampino & Stothers 1984, Clube & Napier 1986). However, there are problems, as we shall now describe.

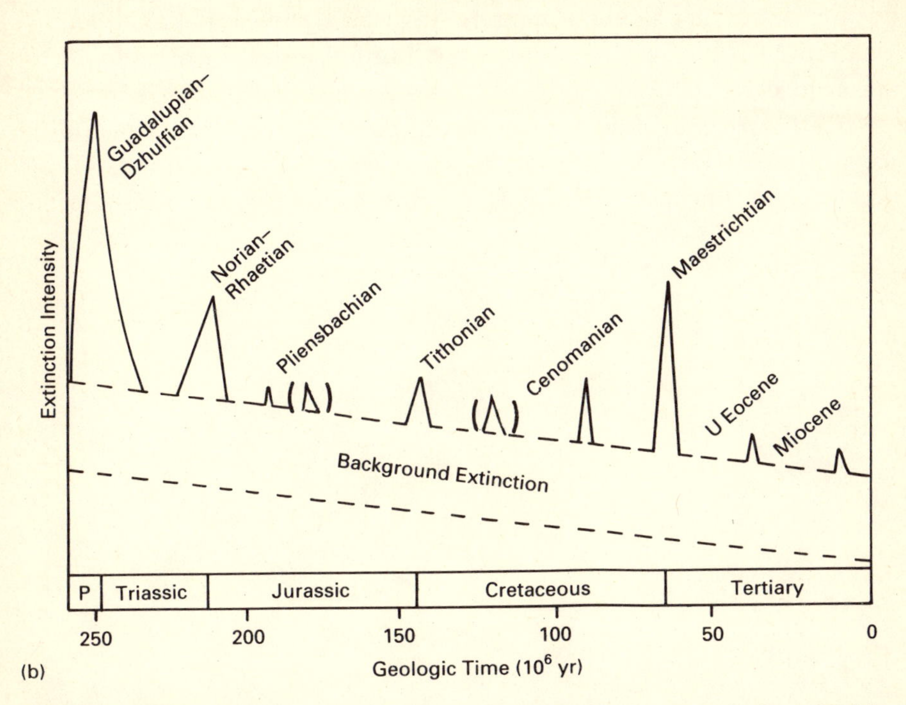

Figure 1(b). The extinction intensity in the marine realm over the last 270 Myr (Sepkoski & Raup 1986). The background extinction rate is shown as a decreasing function of time, while the heights of the eight extinction peaks indicate the relative magnitudes of events at the familial and stage levels. The events shown in brackets were added in later work by Raup & Sepkoski (1988):

The periodic cometary shower hypothesis

First, there is the problem of the validity of the periodicity. Although one can argue about the extent to which the periodicity is well founded on statistical grounds we accept it as being a fact in the present analysis.

Secondly, there is the problem of interactions with GMCs. The hypothesis of GMC interactions causing the release of comets from the Oort cloud is probably the most reasonable. Although it is likely that such interactions do cause comets to be released there are a number of objections to the hypothesis of periodicity, as follows:

(i) The half-width of the molecular cloud layer (70 ± 10 pc) is too large with respect to the amplitude of the Solar orbit (70 ± 20 pc) for the 'signal to noise' ratio to be at all significant. (Thaddeus & Chanan 1985). A further point is the 'lumpy' z-distribution of GMCs; see Figure 2.

(ii) Concerning the mean rate of cometary arrivals, the number density of GMCs is insufficient to give rise to a rate of comets of one per 30 Myr, one per 200-500 Myr being nearer the truth (Bailey *et al.* 1987).

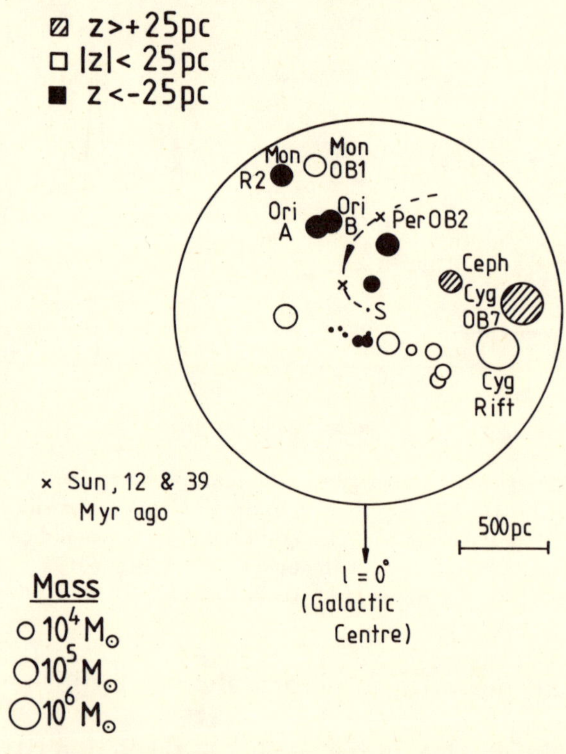

Figure 2. Plan view of the Galaxy showing the present positions and masses of giant molecular clouds within 1 kpc of the sun (Dame *et al.* 1987). The current epicyclic orbit of the sun ($P \simeq 1.7 \times 10^8$yr) is shown, the crosses indicating the positions of the last two mass extinctions (cf. Figure 1):

(iii) Turning to the actual, as distinct from the smoothed, distribution of GMCs, inspection of Figure 2 shows the situation within 1 kpc of the Sun as evinced by the Columbia CO observations (Dame *et al.* 1987).

Figure 2 also shows the projected present epicyclic orbit of the

Solar System (period P $\simeq$ 170 Myr) together with crosses to indicate the Sun's position at the times ($\simeq$ 10 and 40 Myr BP) associated with the last two mass extinctions identified in Figure 1. Insofar as the likely velocities of the GMCs are < 5 km s^{-1} (compared with the current planar velocity of the Solar System of $\simeq$ 18 km s^{-1}) it does not seem likely that the Sun has recently passed close to a GMC, particularly for the 10 Myr BP extinction. Concerning that at $\simeq$ 40 Myr, uncertainties in the cloud motions over this longer interval mean that a past passage close to either Ori A/B or Per OB2 cannot be ruled out. However, even here there is a problem because for an encounter with a molecular cloud to cause a significant comet shower (i.e. one with $a_{sh} < 2 \times 10^4$ AU), the impact parameter for the encounter should satisfy $b < 20\ M_5^{1/3}$pc, where $M_5 = M/10^5 M_\odot$, whence $b < 20$pc for these clouds (Bailey *et al.* 1987). The problem is now that their linear extent is generally greater than this and for an impact parameter of 20 pc from the centre of mass of the cloud, the trajectory will pass through the cloud, with a consequent reduction in the strength of gravitational interaction.

Clube & Napier (1986) have pointed out that the amplitude of the Solar motion may have been greater in the past due to strong interactions with GMCs which may have reduced the z-component of the Solar velocity. However, inspection of Figure 2 shows that such an encounter is not likely to have occurred during the present phase of the Solar oscillation, there being no nearby clouds of sufficient mass to cause the required change of Solar velocity. Moreover, if an earlier strong interaction had occurred, thus altering the phase of the Sun's vertical oscillation, then why is the mean spacing of the last two extinctions the same as the long-term average?

Thus a further problem in connection with this possibility concerns the change in periodicity which would arise from a large change of amplitude. It is well known that the acceleration constant K_z (the equivalent of g) varies considerably and non-linearly with z. Specifically, following Allen (1981), K_z is 2.5, 4.1, 5.8, 6.4 and 7.3, all times 10^{-9} cm s^{-2}, for z = 0.1, 0.2, 0.5, 1.0 and 2.0 kpc, respectively. In order to achieve an acceptable signal to noise ratio we need an amplitude approaching 0.6 kpc and the oscillation period here is some 40% longer than that for the more normal 0.07 kpc. Figure 3 summarises the situation. The result is that the change in amplitude

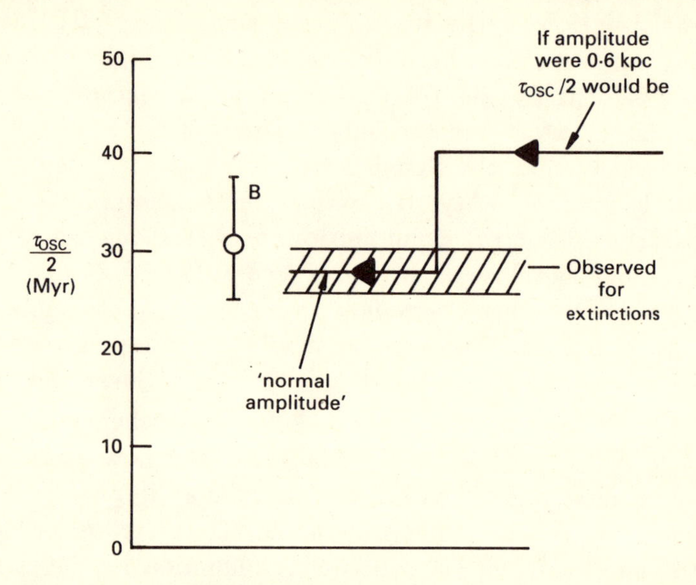

Figure 3. The extinction period (shaded area) and the half-period of the solar oscillation. 'B' denotes the range estimated by Bahcall (1986), assuming no interaction with other bodies. The uncertainty arises because of lack of knowledge of the distribution of dark matter. If the amplitude had been 0.6 kpc at an earlier epoch (sufficient to give an adequate signal to noise ratio for interactions with GMC's) the oscillation period would have been $\simeq$ 40% greater than its more recent value:

would destroy the apparent constancy of period seen in Figure 1. A way out would appear to be to invoke a very recent change but then there are questions: 'where is the massive local perturber?' and, if extinctions are correctly associated with exceptional cometary showers, 'where is the surely super-massive extinction event?'

Thirdly, there is the problem of what happens when the Solar System passes through GMCs or other obnoxious entities. Many workers have invoked the effect on solar irradiance *etc.* of passage through GMCs or indeed gas clouds in general and the attendant implications for terrestrial life. Reference to Figure 4, which shows an enlarged view of the 'local' region, indicates an interesting feature. This is that the last 15 Myr period of transit of the solar system appears to have been through a region of remarkable low gas density (see caption for details). Not only have 'we' been too far from Taurus for its effect to be important (its mass is, in any case, only $\simeq 10^4\ M_\odot$) but there appears to be a deficit of potentially disturbing gas. We are mindful of the fact that this situation - and the shape of the 'hole' in Figure 4 - is probably due to past supernovae, themselves undoubtedly

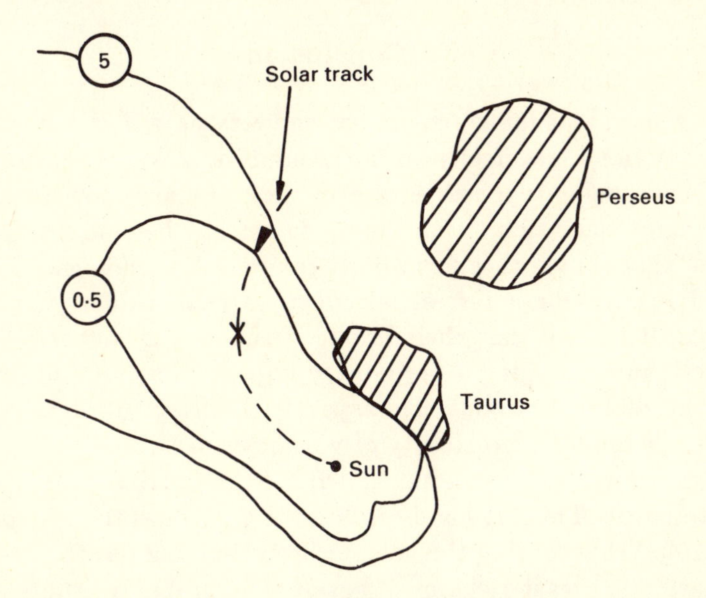

Figure 4. The most recent extinction. This figure represents an enlarged version of part of Figure 2. The track of the sun is marked together with the latter's position at the epoch of the most recent extinction. Also shown are contours of atomic hydrogen in units of $10^{19}\mathrm{cm}^{-2}$ (Frisch & York 1983). It will be noticed that the last extinction was well inside a local low gas density region; an obvious astronomical explanation is therefore unlikely:

generators of dramatic extinctions if close enough, but the chance of such close explosions is vanishingly small (only a few are expected per 4×10^9yr). Fourthly, there is the problem of interactions with other bodies. Bailey *et al.* (1987) derived the range of individual masses of GMC's needed for the periodic disruption hypothesis and their associated surface density. Other known entities having been ruled out, one is thrown back on the ubiquitous dark matter candidates which might be thought to have sufficiently flexible characteristic. However, detailed inspection fails to reveal possible parameters, mainly because of the large scale height apparently needed (Bahcall 1986) for such dark matter entities to provide the appropriate local missing mass.

In fact, we are rather impressed by recent work by Boulares (1988), which casts doubt on the existence of any missing mass in the Solar vicinity.

Conclusion

We have argued against cometary showers as a likely vehicle for mass extinctions specifically on the grounds of no known astronomical phenomenon being reponsible for periodic cometary showers, and this is our main conclusion. Continuing somewhat beyond our brief we presume that some other form of astronomical phenomenon, such as asteroid impacts or a terrestrial effect, is responsible. Concerning asteroids, it is true that their impact frequency is higher than that of comets but arranging a periodicity appears difficult (although we cannot be dogmatic about this aspect). Turning to terrestrial phenomena, there is the possibility of volcanoes providing the necessary ingredients for mass extinctions, with either periodic or aperiodic characteristics. Thus, it has been argued very recently by Courtillot *et al.* (1988) that, during the volcanic events giving rise to the Deccan flood basalts, at least 10^6 km^3 of basalt were probably erupted in less than 1 Myr during the interval 69 to 65 Myr BP. The timing overlaps the K-T boundary (which is without doubt the best documented and most dramatic extinction) and there is thus the implication that the extinctions could have been caused by the world-wide effects associated with this volcanism. The question then arises as to the cause of the great volcanic events - it would be a strange quirk if they themselves were initiated by comet or asteroid impacts, as suggested very recently by Rampino & Stothers (1988).

References

Allen C.W. 1981 ASTROPHYSICAL QUANTITIES. (Athlone Press, London).

Alvarez L.W., Alvarez W., Asarao F. & Michel H.V. 1980 *Extraterrestrial cause for the Cretaceous-Tertiary extinction.* Science **208**, 1095.

Bahcall N.J. 1986 *The galactic environment of the Solar System.* THE GALAXY AND THE SOLAR SYSTEM. (eds Smoluchowski R., Bahcall

J.N. & Matthews M.S., University of Arizona Press, Tucson), page 3.

Bailey M.E., Wilkinson D.A. & Wolfendale A.W. 1987 *Can episodic comet showers explain the 30 Myr cyclicity in the terrestrial record?* Mon. Not. R. astr. Soc., **227**., 863.

Boulares A. 1988 Private Communication

Clube S.V.M. 1978 *Does the Galaxy have a violent history?* Vistas Astr. **22**, 77.

Clube S.V.M. & Napier W.M. 1986 *Giant comets and the Galaxy: implications of the terrestrial record.* THE GALAXY AND THE SOLAR SYSTEM. (eds. Smoluchowski R., Bahcall J.N. & Matthews M.S., University of Arizona Press, Tucson), page 260.

Courtillot V., Feraud G., Maluski H., Vandamme D., Moreau M.G. & Besses J. 1988 *Deccan flood basalts and the Cretaceous- Tertiary boundary.* Nature **333**, 843.

Dame T.M., Ungerechts H., Cohen R.S., de Gues E.J., Grenier A., May J., Murphy D.C., Nymn L-A. & Thaddeus, P. 1987 *A composite CO survey of the entire Milky Way.* Astrophys. J. **322**, 706.

Frisch P.C. & York, D.G. 1983 *Synthesis maps of ultraviolet observations of neutral interstellar gas.* Astrophys. J. **271**, L59.

Napier W.M. & Clube S.V.M. 1979 *A theory of terrestrial catastrophism.* Nature **282**, 455.

Rampino M.R. & Stothers R.B. 1984 *Terrestrial mass extinction; cometary impact and the Sun's motion perpendicular to the Galactic plane.* Nature **308**, 709.

Rampino M.R. & Stothers R.B. 1988 Private Communication

Raup D.M. & Sepkoski J.J. 1984 *Periodicity of extinctions in the geologic past.* Proc. Natl. Acad. Sci. **301**, 801.

Sepkoski J.J. & Raup D.M. 1986 DYNAMICS OF EXTINCTION. (ed. Elliot D.K., Wiley-Interscience, New York), page 3.

Thaddeus P. & Chanan G.A. 1985 *Cometary impacts, molecular clouds, and the motion of the Sun perpendicular to the Galactic plane.* Nature **314**, 73.